T0073046

SCIENCE, MUSIC, AND MATHEMATICS

The Deepest Connections

Second Edition

SCIENCE, MUSIC, AND MATHEMATICS
The Deepest Connections

Second Edition

Michael Edgeworth McIntyre
University of Cambridge, UK

World Scientific

NEW JERSEY · LONDON · SINGAPORE · BEIJING · SHANGHAI · HONG KONG · TAIPEI · CHENNAI · TOKYO

Published by

World Scientific Publishing Co. Pte. Ltd.

5 Toh Tuck Link, Singapore 596224

USA office: 27 Warren Street, Suite 401-402, Hackensack, NJ 07601

UK office: 57 Shelton Street, Covent Garden, London WC2H 9HE

Library of Congress Cataloging-in-Publication Data
Names: McIntyre, Michael Edgeworth, 1941–   author.
Title: Science, music, and mathematics : the deepest connections /
    Michael Edgeworth McIntyre, University of Cambridge, UK.
Description: Second edition. | New Jersey : World Scientific, [2023] |
    Includes bibliographical references and index.
Identifiers: LCCN 2023021457 | ISBN 9789811276972 (hardcover) |
    ISBN 9789811278501 (paperback) | ISBN 9789811276989 (ebook for institutions) |
    ISBN 9789811276996 (ebook for individuals)
Subjects: LCSH: Science and music. | Climate change. | Music--Mathematics.
Classification: LCC ML3800 .M295 2023 | DDC 781.1--dc23/eng/20230517
LC record available at https://lccn.loc.gov/2023021457

British Library Cataloguing-in-Publication Data
A catalogue record for this book is available from the British Library.

For the book's supplementary material, please visit
https://www.worldscientific.com/worldscibooks/10.1142/13429#t=suppl

Desk Editors: Nambirajan Karuppiah/Ana Ovey

Typeset by Stallion Press
Email: enquiries@stallionpress.com

The first edition of Michael Edgeworth McIntyre's book rightly received plaudits for its insightfulness, clarity of thought, unique approach, and explanation of cross-cutting topics in science, music and mathematics. One of its highlights, for me as an atmospheric physicist, was the Postlude "The amplifier metaphor for climate" which discusses how the climate system can respond significantly to small changes in inputs, such as solar energy, and how these are magnified by the greater availability of "fuel" in the form of water vapour associated with human-produced global warming. In the second edition this discussion is augmented by sections on the evidenced increase in occurrence of extreme weather events and the physics behind the possibility of climate tipping points. A key theme of the book is the clear-headed objectivity needed in communicating complex scientific results to non-specialists, both to support policymakers in acting on climate change and to call out those who try to distort such information. Michael's writing exemplifies how such clarity can be achieved without detracting from a really good read.

— Professor Joanna Haigh CBE FRS,
Imperial College London

Michael Edgeworth McIntyre is one of the world leading researchers on the fluid dynamics of our climate system. As such he is an expert on the mathematical nature of our home, Planet Earth. But remarkably, were it not for the flap of a butterfly's wings, he may have become a professional classical violinist. In this remarkable book, McIntyre weaves together these three themes – science, music and mathematics – into a unified whole. The result is a lucid and unique discussion, buzzing with infectious enthusiasm, of the deep interrelationships between the arts and the sciences. The book should appeal to anyone interested in what makes us the creative species we are.

— Professor Timothy Palmer CBE FRS,
University of Oxford, author of
The Primacy of Doubt

This book superbly reflects its author in its intellectual depth and breadth, and in its clarity and enthusiasm. It is a book to be read by all, but particularly by young scientists, as it gives a clarion call for them to play their vital role in the quest to help our species thrive in a sustainable manner on this planet.

— Professor Sir Brian Hoskins CBE FRS,
Grantham Institute for Climate Change

Contents

Foreword

by Herbert E. Huppert FRS

Michael Edgeworth McIntyre FRS has written a book wide in its intellectual range and provocative in its implications. That will surprise no-one who knows him. His deep yet expansive vision of the world goes back to his ancestry and upbringing. His father, Archie McIntyre FAA, was a respected neurophysiologist and his mother, Anne McIntyre, was an accomplished visual artist. His great-grandfather, Sir Edgeworth David FRS, was a famous geologist and Antarctic explorer.

Michael's childhood experiences, in Australia and New Zealand, included observing his father's experimental skill, dissecting out individual nerve fibres in a laboratory full of electronic amplifiers and oscilloscopes. That was in the early days of micro-electrodes able to record from a single nerve cell. Michael also remembers being fascinated as a small boy by the sound of Beethoven's Eighth Symphony made visible on an oscilloscope. He became very curious about how that sound — the sound of the "marvellously life-enhancing, energetic first movement", as he described it — was related, somehow, to the wiggling green line on the oscilloscope tube.

In his teens Michael became a skilled violinist, and in 1960 rose to be leader of the New Zealand National Youth Orchestra. Such was his talent and his love of music that, after coming to Cambridge to work for a doctorate in fluid dynamics, he considered becoming a full-time professional musician — and indeed did become a part-time professional, and a member of the British Musicians' Union, for some years. However, to our great good fortune, mathematics and science exerted a stronger pull in the end. He continues to play violin and viola and to passionately explore the links of music to other fields, notably neurology.

The book is divided into sections, the anecdotal narratives dancing gracefully with the scholarly enquiries, explicit or implicit, around the opening question "What is science?"

Rarely in this era of super-specialization does one encounter a true polymath, but that is a fair title to bestow on Michael. He has made original contributions to the fundamental understanding of fluid dynamics in the Earth's atmosphere, in its oceans and in the Sun's interior. He writes persuasively about the importance of lucidity and "lucidity principles", whether in writing and speaking, or in designing a nuclear reactor (and everything in between). He is concerned about the unconscious assumptions that impede human progress, not least those stemming from our evolutionarily ancient dualism or "dichotomization instinct", which "makes us stupid" in the manner now so dangerously amplified by the social media, through their "artificial intelligences that aren't yet very intelligent". But we can still hope that "smarter robots could be a game-changer in helping us humans to get smarter, too." Michael argues for the importance of scientific ethics, and reminds us of the limits of our epistemologies.

Michael is an engaging companion on this highly informed but charmingly informal odyssey. He invites the reader to see new landscapes through new lenses, and, crucially, to make new connections in all sorts of ways.

> *"There are more things in heaven and earth, Horatio, than are dreamt of in your philosophy."*

Preface to the second edition

This book brings together a set of interconnected insights from the sciences and the arts that are so deep, and so basic, I think, in so many ways, that they deserve to be better known than they are. They include insights into communication skills, and into the power and limitations of science. They are 'deep' in the sense of being deep in our nature, and evolutionarily ancient. I go on to discuss, at the end, what science can and can't tell us about the climate problem.

So I hope that the book will interest scientifically minded readers, especially young scientists. I've tried to bring out the connections in a widely understandable way that avoids equations and technical jargon. The book draws on a lifetime's personal experience as a scientist, as a mathematician, and as a musician.

It hardly needs saying that if you're a young scientist you need to hone your communication skills. In our complex world such skills are needed not only between scientists and the public, but also between scientists in different specialities. I'll argue that it's useful to know, for instance, how the skilful use of written and spoken language can be informed by the way music works — music of any genre.

Good communication will be crucial to tackling today's and tomorrow's great problems, including the problems of climate change, biodiversity, and future pandemics, and, now more urgently than ever, the problem of understanding the strengths and weaknesses of artificial intelligence. One obstacle to understanding is the misconception — it's sometimes called the 'singularity fallacy' — that there's only one kind of intelligence, and only one measure of intelligence.

To develop our understanding of how communication works, it's useful and interesting to consider the origins of human language. It's now known that biological evolution gave rise to our language ability in a manner quite different from what popular culture says. And evolution is itself now better understood. Rather than a simple competition between selfish genes, it's more like turbulent fluid flow, in some ways — a complex process spanning a vast range of timescales. I try to discuss those points very carefully, since they're still contentious.

A further aspect of evolution is that, again contrary to popular belief, there's a genetic basis not only for the nastiest but also for the most compassionate, most cooperative parts of human nature.

Throughout the book I've aimed for high-quality reasoning and respect for evidence. I've tried to keep the main text as short as possible, and readable without reference to the voluminous endnotes. I recommend ignoring the endnotes on a first read. However, as well as adding to the arguments, the endnotes give what I hope are enough literature references to support what I say. The literature is moving fast; and this second edition incorporates a number of significant updates. One of them is a truly remarkable new book[118] by my climate-science colleague Tim Palmer FRS.

Note regarding video and audio clips: Figure 1 below presents an animated video clip that is basic to many of my arguments.

The QR code in the figure points to a site on YouTube that should display the video clip on any smartphone with a QR reader. Alternatively, that clip, together with some audio clips from later figures, can be downloaded from the World Scientific website as a zip file. The URL is https://www.worldscientific.com/worldscibooks/10.1142/13429#t=suppl.

Acknowledgements: Many kind friends and colleagues have helped me with advice, encouragement, information, and critical comments over the years. I have learnt much from experts on the latest developments in systems biology, evolutionary theory, and palaeoclimatology. Beyond those mentioned in the acknowledgements sections and endnotes of my original *Lucidity and Science* papers,[34, 75, 130] I'd like to thank Dorian Abbot, Leslie Aiello, David Andrews, Paul Ashworth, Grigory Barenblatt, Terry Barker, George Batchelor, Pat Bateson, Liz Bentley, Francis Bretherton, Oliver Bühler, Frances Cairncross, David Crighton, Ian Cross, Judith Curry, Philip Dawid, David Dritschel, Kuniyoshi Ebina, George Ellis, Sue Eltringham, Kerry Emanuel, Matthew England, David Fahey, Georgina Ferry, Rupert Ford, Angela Fritz, Chris Garrett, Jeffrey Ginn, Douglas Gough, Richard Gregory, Stephen Griffiths, Joanna Haigh, Peter Haynes, Isamu Hirota, Brian Hoskins, Matthew Huber, Herbert Huppert, James Jackson, Sue Jackson, Philip Jones, Peter Killworth, Kevin Laland, Steve Lay, James Lighthill, Zheng Lin, Paul Linden, Shyeh Tjing Loi, Malcolm Longair, Jianhua Lü, James Maas, David MacKay, Normand MacLaurin, Niall Mansfield, David Marr, Nick McCave, Evelyn McFadden, Amy McGuire, Richard McIntyre, Ruth McIntyre, Steve Merrick, Gos Micklem, Alison Ming, Simon Mitton, Ali Mohebalhojeh, Ken Moody, Brian Moore, Walter Munk, Alice Oven, Tim Palmer, Antony Pay, Anthony Pearson, Tim Pedley, Sam Pegler, Max Perutz, Ray Pierrehumbert, Miriam Pollak, Vilayanur

Ramachandran, Dan Rothman, Murry Salby, Adam Scaife, Nick Shackleton, Ted Shepherd, Adrian Simmons, Bill Simmons, Emily Shuckburgh, Luke Skinner, Appy Sluijs, David Spiegelhalter, Marilyn Strathern, Zoe Strimpel, Daphne Sulston, John Sulston, Stephen Thomson, Paul Valdes, Yixin Wan, Andy Watson, Ronald S. Watts, Estelle Wolfers, Flick Wolfers, Jeremy Wolfers, Jon Wolfers, Peter Wolfers, Eric Wolff, Toby Wood, Jim Woodhouse, and Laure Zanna. And I owe two very special debts of gratitude. One is to the clarinettist and conductor Antony Pay, a man of profound insight not only into the workings of music but also into science, humanity, and human nature. At a late stage he kindly read the entire manuscript and made many helpful suggestions. My other special debt is to Alice Oven, without whom this book would never have been written. As a young commissioning editor for World Scientific, it was she who first got me interested in embarking on this project, building on Refs. 130. For valuable help in seeing the First Edition through to press I'm grateful to World Scientific's Swee Cheng Lim, Jing Wen Soh, and Rok Ting Tan, and for the Second Edition Ana Ovey and Nambirajan Karuppiah.

— MEM
Cambridge, February 2023

Key phrases: *science, communication skills, language, lucidity principles, organic-change principle, music, mathematics, human evolution, selfish-gene metaphor, false dichotomies, nature versus nurture, dichotomization instinct, disinformation, social-media weaponization, artificial intelligence, singularity fallacy, perception psychology, time perception, acausality illusions, out-of-body experiences, climate change, weather extremes, weather fuel*

About the author

Professor Michael Edgeworth McIntyre is an eminent scientist who has also had a part-time career as a musician. In this book he offers an extraordinary synthesis, revealing the many deep connections between science, music, and mathematics. He avoids equations and technical jargon. The connections are deep in the sense of being embedded in our very nature, rooted in biological evolution over hundreds of millions of years.

Michael guides us through biological evolution, perception psychology, and even unconscious science and mathematics, all the way to the scientific uncertainties about the climate crisis.

He also has a message of hope for the future. Contrary to popular belief, he holds that biological evolution has given us not only the nastiest, but also the most compassionate and cooperative parts of human nature. This insight comes from recognizing that biological evolution is far more than a simple competition between selfish genes. Instead, he argues, in some ways it is more like the turbulent, eddying flow in a river or in an atmospheric jet stream, a complex process spanning a vast range of timescales.

Professor McIntyre is a Fellow of the Royal Society of London (FRS) and has long been interested in how different branches of science can better communicate with each other, and with the public. His work harnesses aspects of neuroscience and psychology that point toward the deep 'lucidity principles' that underlie skilful communication, principles related to the way music works — music of any genre.

This Second Edition sharpens the previous discussion of communication skills and their importance for today's great problems, ranging from the widely discussed climate crisis to the need to understand the strengths and weaknesses of artificial intelligence.

1

The unconscious brain

Consider for a moment the following questions.

1. What is science? What is music? What is mathematics? Are there connections between music and mathematics that go deeper than the usual games with numbers?

2. What is lucidity — of writing, speaking, thinking, and design? How can one branch of knowledge best communicate with another? How can we navigate the vast conceptual minefield known as human language?

3. Where does our sense of truth, beauty, and transcendence come from? What is the Platonic world of perfect mathematical forms? Is our Universe part of that world, as some physicists think?

4. Why are people instinctively hypercredulous — driven to believe in an Absolute Truth or Answer to Everything, regardless of evidence, not only within the fundamentalist religions and atheisms but also, surprisingly, sometimes even among scientists?

5. Why do people instinctively dichotomize or polarize things, regardless of evidence? What *are* those perilous binary buttons that demagogues and others keep on pressing?

6. And what indeed are these things called instincts? What kind of thing is genetic memory? Have the genetic-blueprint and selfish-gene metaphors been misleading us in any way?
7. How can we best understand the power and limitations of science? And what can science tell us about, for instance, the climate problem and its uncertainties?

Good answers are important to our hopes of a civilized future; and many of the answers are surprisingly simple. But a quest to find them will soon encounter a conceptual and linguistic minefield, some of it around ideas like 'innateness' and 'instinct'. Still, I think I can put us within reach of some good answers (with a small 'a') by recalling, first, some points about how our pre-human ancestors must have evolved — in a way that differs crucially from what popular culture says — and, second, some points about how we perceive and understand the world.

One reason for looking at evolution is the prevalence of misconceptions about it. Chief among them is the idea that natural selection works solely by competition between individuals. That ignores many examples of cooperative behaviour among social animals, some of which Charles Darwin himself was at pains to point out.[1] Saying that competition between individuals is all that matters flies in the face of this and much other evidence. It has also done great damage to human societies.[2]

On perception, understanding, and misunderstanding, and on our extraordinary language ability, it hardly needs saying that they were shaped by our ancestors' evolution. Less obvious, however, is that the evolution must have depended not only on cooperation alongside competition but also, according to the best evidence, on a powerful feedback between genomic evolution and cultural evolution. For instance, our language ability couldn't have been suddenly invented around a hundred millennia ago, purely as a result of cultural evolution, as some researchers have argued. On the contrary, I'll show in Chapter 3 — drawing on clinching evidence from Nicaragua — that our language ability must have developed much more gradually, through the co-evolution of genomes and cultures with each affecting the other over a much longer timespan, probably millions of years.

Such co-evolution is necessarily a *multi-timescale process*. Multi-timescale processes are ubiquitous in the natural world. They're found everywhere. They depend on strong feedbacks between different mechanisms over a large range of timescales. Here we have slow and fast mechanisms in the form of genomic evolution and cultural evolution. I'm using the word 'cultural' in a broad sense, to include everything passed on by social learning. Such feedbacks have often been neglected in the literature on biological evolution. Their likely importance for pre-human evolution and their multi-timescale aspects were, however, recognized and pointed out as long ago as 1971 by the great biologist Jacques Monod,[3] and by the great palaeoanthropologist Phillip Tobias.[4]

Another theme in this book will be unconscious assumptions. It's clear that such assumptions underlie, for instance, the polarized debates about 'nature versus nurture', 'instinct versus learning', 'genomic evolution versus cultural evolution', and so on. A gut feeling that evolution is either genomic or cultural, *with each excluding the other*, is typical. At a deeply unconscious level, it's assumed that you can't have both together. Further examples will come up in Chapter 3. They're germane to some notable scientific controversies.

The dichotomization instinct, as I'll call it, the visceral push toward polarization — toward seeing all choices and distinctions as binary and exclusive — is by no means the only source of unconscious assumptions. Much more of what's involved in perception and understanding, and in our general functioning, takes place unconsciously. Some people find this hard to accept. Perhaps they feel offended, in a personal way, to be told that the slightest aspect of their existence might, just possibly, not be under full and rigorous conscious control. A brilliant scientist whom I know personally as a colleague took offence in exactly that way, in a discussion we had on unconscious assumptions in science — even though the exposure of such assumptions is the usual way in which scientific knowledge improves, as history shows again and again, and even though I offered clear examples from our shared field of expertise.[5]

Many other examples are given in the book by Daniel Kahneman.[6] My own favourite example is a very simple one, Gunnar Johansson's 'walking dots' or 'walking lights' animation. Twelve moving dots in a two-dimensional plane are unconsciously assumed to represent a particular three-dimensional motion. When the dots are moving, everyone with normal vision sees a person walking. To see the animation, point your smartphone at the QR code on the right of Figure 1; in case that fails to work, see the alternatives in the figure caption:

Figure 1: On the left is a single frame from Gunnar Johansson's 'walking dots' animation. On the right is a QR code that should display the animation on a smartphone with a QR reader, via YouTube. (In some e-readers you can also click on the QR code; and in case none of this works, the animation, plus other clips from later figures, can be downloaded as a zip file from https://www. worldscientific.com/worldscibooks/10.1142/13429#t=suppl.) The animation shows a person walking from far right to near left. The walking dots phenomenon is a well-studied classic in experimental psychology and is one of the most robust perceptual phenomena known. Animation constructed by Steve Lay from data kindly supplied by Professor James Maas.

And again, anyone who has driven cars, or flown aircraft, will probably remember occasions on which accidents were avoided ahead of conscious thought. The typical experience is often described as *witnessing oneself* taking, for instance, evasive action when faced with a head-on collision, or other life-threatening emergency. It is all over by the time conscious thinking has begun. It has happened to me, in cars and in gliders. I think such experiences are quite common. Kahneman gives an example from firefighting.[6]

Many years ago, the anthropologist-philosopher Gregory Bateson put the essential point succinctly, in classic evolutionary cost–benefit terms[7]:

"No organism can afford to be conscious of matters with which it could deal at unconscious levels."

Gregory Bateson's point applies to us as well as to other living organisms. Why? There's a mathematical reason, combinatorial largeness. Every living organism has to deal all the time with a combinatorial tree, a combinatorially large number, of present and future possibilities. Each branching of possibilities multiplies, rather than adds to, the number of possibilities. Being conscious of all those possibilities would be almost infinitely costly.

Combinatorially large numbers are *unimaginably* large. No-one can feel their magnitudes intuitively. For instance, the number of ways to shuffle a pack of 52 cards is $52 \times 51 \times 50 \times \ldots \times 3 \times 2 \times 1$. That's just over eighty million trillion trillion trillion trillion trillion.

The 'instinctive' avoidance of head-on collision in a car — the action taken ahead of conscious thought — is not, of course, something that comes exclusively from genetic memory. Learning is involved as well. The same goes for the way we see the walking dots animation. But much of that learning is itself unconscious, stretching back to the infantile groping that discovers the outside world and allows normal vision to develop.[8] Far from being mutually exclusive, nature and nurture are intimately intertwined. That intimacy stretches even further back, to the genome within the embryo 'discovering' and interacting with its maternal environment.[9] *Jurassic Park* is a great story, but scientifically wrong because you need dinosaur eggs as well as dinosaur DNA. Who knows, though — since birds are dinosaurs someone might manage it, one day, with reconstructed DNA and birds' eggs.

My approach to questions like the foregoing comes from long experience as a scientist. Science was my main profession for fifty years or so. Although many branches of science interest me, my professional career was focused mainly on mathematical research to understand the highly complex, multi-timescale fluid dynamics of the Earth's atmosphere and oceans. Included are phenomena such as the great jet streams and the air motion that shapes the ozone hole in the Antarctic stratosphere, and what are sometimes called the "world's largest breaking waves". Imagine a giant sideways breaker in the stratosphere the mere tip of which is almost as large as the entire USA. That research has in turn helped us, in an unexpected way, to understand the complex fluid dynamics and magnetic fields of something even more gigantic, the Sun's interior. But long ago I almost became a musician. Or rather, in my youth I was, in fact, a part-time professional musician and could have made it into a full-time career. So I've had artistic preoccupations too, and artistic aspirations. This book tries to get at the deepest connections between *all* these things.

It's obvious, isn't it, that science, mathematics, and the arts are all of them bound up with the way perception works. That'll be the central topic in Chapter 4, where the walking dots will prove informative. And common to science, mathematics, and the arts is the creativity that leads to new understanding, the thrill of curiosity and lateral thinking, and sheer wonder at the whole phenomenon of life itself and at the astonishing Universe we live in.

One of the greatest of those wonders is our own adaptability, our versatility. Who knows, it might even get us through today's crises, desperate though they might seem. We know that our hunter-gatherer ancestors were highly adaptable. They were driven again and again to migration and different ways of living by, among other things, rapid climate fluctuations — the legendary years of famine and years of plenty. How else did our species — a single, genetically compatible species with its single human genome — spread around the globe in less than a hundred millennia? Chapter 3 will point to recent hard evidence for the sheer rapidity, and magnitude, of some of those climate fluctuations.

Chapter 3 will also point to recent advances in our understanding of biological evolution and natural selection, advances not yet assimilated into popular culture. One implication is that not only the nastiest but also the most compassionate, most cooperative parts of our makeup are 'biological' and deep-seated.[2, 10, 11] There's a popular misconception — yet another variation on the theme of nature 'versus' nurture — that our nastiest traits are exclusively biological, and our nicest traits exclusively cultural. We'll see that the evidence says otherwise.

Here, by the way, as in most of this book, I lay no claim to originality. For instance the evidence on past climates comes from the painstaking work of colleagues at the cutting edge of palaeoclimatology, including great scientists such as the late Nick Shackleton whom I had the privilege of knowing personally. And the points I'll make about biological evolution rely on insights gleaned from colleagues at the cutting edge of biology, including the late John Sulston of human-genome fame, whom I also knew personally.

Our ancestors must have had not only language and lateral thinking — and music, dance, poetry, and storytelling — but also rhetoric, power games, blame games, genocide, ecstatic suicide, and the rest. To survive, they must have had love and compassion too. The precise timespans and evolutionary pathways for these things are uncertain. But the timespans for at least some of them, including the beginnings of our language ability, must have been a million years or more to allow for the multi-timescale co-evolution of genomes and cultures.

As already suggested there's been a tendency to neglect such co-evolution despite the ubiquity — the commonplace occurrence — of other multi-timescale processes in the natural world. Of these there's a huge variety. To take one of the simplest examples, consider air pressure, as when pumping up a bicycle tyre. Fast molecular collisions mediate slow changes in air pressure, and air temperature, while pressure and temperature react back on collision rates and strengths. That's a strong and crucial feedback across enormously different timescales.

So it never made sense to me to say that long and short timescales can't interact. It never made sense to say that genomic evolution can have no interplay with cultural evolution just because the one is slow and the other is fast. And in particular it never made sense to argue from the archaeological record, as some researchers have, that language started around a hundred millennia ago as a purely cultural invention — the sudden invention of a single mother tongue from which today's languages are all descended, purely by cultural transmission.[12, 13] I'll return to these points in Chapter 3 and will try to argue them very carefully.

When considering the archaeological record, it's sometimes forgotten that language and culture can be mediated purely by sound waves and light waves and held in individuals' memories — as in the *Odyssey* or in a multitude of other oral traditions, including Australian aboriginal songlines, the Japanese epic *Tale of the Heike*, and the many stories in what Laurens van der Post has called the immense wealth of the unwritten literature of Africa.[14] That's a very convenient, and eminently portable, form of culture for a tribe on the move. And sound waves and light waves are such ephemeral things. They have the annoying property of leaving no archaeological trace. But absence of evidence isn't evidence of absence.

And now, in a mere flash of evolutionary time, a mere few centuries, we've shown our versatility and adaptability in ways that seem to me more astonishing than ever. We no longer panic at the sight of a comet. Demons in the air have shrunk to a small minority of alien abductors. We don't burn witches and heretics, at least not literally. The Pope apologizes for past misdeeds. Genocide was avoided in South Africa. We even dare, sometimes, to tolerate individual propensities and lifestyles if they don't harm others. We argue that tyrants needn't always win. And guess what, they *don't* always win. Indeed, recent reversals notwithstanding, governments have become less tyrannical and more democratic, on average, over the past two centuries, in what political scientist Samuel Huntington has called three waves of democratization[15] despite the setbacks in between, and now. And most astonishing of all, since 1945 we've even had the good sense so far — and very much against the odds — to avoid warfare with nuclear weapons.

We've marvelled at the sight of our beautiful Earth poised above the lunar horizon. We have space-based observing systems, super-accurate clocks, and super-accurate global positioning, adding to the cross-checks on Einstein's gravitational theory, also called general relativity. And now there's yet another, very beautiful cross-check on the theory — detection of the lightspeed gravitational ripples it predicts.[16] We have the Internet, bringing us new degrees of freedom and profligacy of information and misinformation. It presents us with new challenges to exercise critical thinking and to build computational systems and artificial intelligences of unprecedented power, and to use them for civilized purposes, exploiting the robustness and reliability growing out of the open-source software movement, "the collective IQ of thousands of individuals".[17] We can read and write genetic codes, and thanks to our collective IQ are beginning, just beginning, to understand them.[18] On large and small scales we've been carrying out extraordinary new social experiments, with labels like 'market democracy', 'market autocracy', 'children's democracy'[19], 'microlending' conducive to population control,[20] 'citizen science', and the burgeoning social media. With the weaponization of the social media now upon us — and the threats to democracy, privacy, and safety from, for instance, automated face recognition, reverse-image search, and deep-fake software[21] — there's a huge downside as with any new technology. But there's also a huge upside, and everything to play for.

An apocryphal story about the Mahatma Gandhi says, in various versions, that he was asked what he thought of modern, or British, or Western civilization. Gandhi is said to have replied, "That would be a good idea." I certainly think that civilization, as distinct from claims to its ownership, would be a good idea. The optimist in me hopes you agree. Part of it would be not only critical thinking and a wider recognition of the power and the limitations of science — including its power, and its limitations, in helping us to understand our own human nature — but also a further healing of the estrangement between science and the arts and humanities, what used to be called the 'two cultures' problem. And, who knows, we might even manage to deepen our understanding of the origin not only of artistic, but also of religious, experience — something that's undoubtedly real and important to many people just as, for me personally, musical experience is real and important. In any case, we need a deeper understanding of our own nature and its diversity and a wider dissemination of such understanding, if only to promote a clear-eyed view of its covert exploitation, and would-be monopolization, by the plutocracies and social media and by the commercial trollers and disinformers, as well as by the political cyberwarfare and conspiracy-theory machines and by the demagogues trying, as always, to replace democracy by autocracy.[21] We need insight into how those agents manipulate our unconscious assumptions.

In trying to get at the deepest connections between these things, I've found that some of the connections can be seen not only as deep but also as simple — provided that one is willing to maintain a certain humility, and willing to think on more than one level.

Multi-level thinking is nothing new. It has long been recognized, unconsciously at least, as being essential to science. It goes back in time before Newton, Galileo, Ibn Sina, and Archimedes. What's complex at one level can be simple at another. Newton treated the Earth as a point mass. It led to his breakthrough in understanding orbital dynamics — despite the enormous complexity of the real Earth.

Today we have a new conceptual framework, complexity theory or complex-systems theory, working alongside the powerful mathematics of the new Bayesian causality theory.[22] The causality theory is now a crucial part of cutting-edge science. It's also part of the secret 'large hadron collider of experimental psychology', as I'll call it, built by the social-media technocrats.[23] In science, the new conceptual framework is helping to clarify what's involved in multi-level thinking and to develop it more systematically, more generally, and more consciously. Key ideas include self-organization, self-assembling components or building blocks, and the use of the Bayesian probabilistic 'do' operator[22] to distinguish correlation from causality. Self-assembling building blocks are also called self-organizing autonomous components or, for brevity, *automata*.

Another key idea is that of *emergent properties* — at different levels of description within complex systems and hierarchies of systems, not least the human brain itself. An emergent property is something that looks simple at one level even though caused by the interplay of complex, often chaotic, events at lower levels. A related idea is that of 'order emerging from chaos'.

We'll see that the ideas of multi-level thinking, automata, and emergent properties are all crucial to making sense of many basic phenomena, such as the way genetic memory works and what instincts are — instincts in the everyday sense referring to things we do, and perceive, and feel automatically, ahead of conscious thought. And without those ideas, including multi-level thinking, there's no chance of making the slightest sense of such things as 'consciousness' and 'free will' — aspects of which I'll touch on in Chapters 4 and 5.

Scientific progress has *always* been about finding a level of description and a viewpoint, or viewpoints, from which something at first sight hopelessly complex can be seen as simple enough to be understandable. The great biologist Peter Medawar called it the art of the soluble. The Antarctic ozone hole is a case in point. I myself made a contribution by spotting some simplifying features in the fluid dynamics, in the way the air moves and transports chemicals. And, by the way, so high is our scientific confidence in today's understanding of the ozone hole — with many observational and theoretical cross-checks — that the professional disinformers who tried to discredit that understanding in a well-known series of campaigns[24] are no longer taken seriously. That's despite the enormous complexity of the problem, to be touched on in Chapter 3, involving spatial scales from the planetary down to the atomic, and multiple timescales from many decades down to thousand-trillionths of a second.

We now have practical certainty, and wide acceptance, that man-made chemicals are the main cause of the ozone hole. We understand in detail why the ozone hole appeared in the south, even though the chemicals causing it were emitted mostly in the north. The chemicals were transported from north to south by global-scale air motion, in a pattern of horizontal and vertical motion whose fluid dynamics was a complete mystery when I began my research career, but is now well understood. Waves and turbulence are involved, in a complex set of multi-timescale processes. And now, through what's called the Montreal Protocol, we have internationally-agreed regulations to restrict emissions of the chemicals despite the disinformers' aim of stopping any such regulation. We have a new symbiosis between market forces and regulation.[25]

What makes life as a scientist worth living? For me, part of the answer is the joy of being honest.

There's a scientific ideal and a scientific ethic that power good science. And they depend crucially on openness and honesty. If you stand up in front of a large conference and say of your favourite theory "I was wrong", you *gain* respect rather than losing it. I've seen it happen. Your reputation *increases*. Why? The scientific ideal says that respect for evidence, for theoretical coherence and self-consistency, for cross-checking, for finding mistakes, for dealing with uncertainty and for improving our collective knowledge is more important than personal ego, or financial gain. And if someone else has found evidence that refutes your theory, then the scientific ethic requires you to say so publicly. The ethic says that you must not only be factually honest but must also give due credit to others, by name, whenever their contributions are relevant.

The scientific ideal and ethic are powerful because even when, as is inevitable, they're followed only imperfectly, they encourage not only a healthy scepticism but also a healthy mixture of competition and cooperation. Just as in the open-source software community, the ideal and ethic harness the collective IQ, the collective brainpower, of large research communities in ways that can transcend even the power of individual greed and financial gain. The ozone-hole story is a case in point.

So too is the human-genome story with its promise of future scientific breakthroughs, including medical breakthroughs, calling for years and decades of collective research effort. The scientific ideal and ethic were powerful enough to keep the genomic information in the public domain — available for use in open research communities — despite an attempt to lock it up commercially that very nearly succeeded.[26] Our collective brainpower will be crucial to solving the problems posed by the genome and the molecular-biological systems of which it forms a part, including the interplay with current and future pandemic diseases. Like so many other problems now confronting us, they are problems of the most formidable complexity.

In the Postlude, I'll return to the struggle between open science and the forces ranged against it, with particular reference to climate change, the most complex problem of them all. Again, there's no claim to originality here. I merely aim to pick out, from the morass of confusion and misinformation surrounding the topic,[24] some basic points clarifying *where* the uncertainties lie, as well as the near-certainties.

2
What is lucidity?
What is understanding?

This book reflects my own journey towards the frontiers of human self-understanding. Of course, many others have made such journeys. But in my case the journey began in a slightly unusual way.

Music and the arts were always part of my life. Music was pure magic to me as a small child. But the conscious journey began with a puzzle. While reading my students' doctoral thesis drafts, and working as a scientific journal editor, managing the peer review of colleagues' research papers, I began to wonder why lucidity, or clarity — in writing and speaking, as well as in thinking — is often found difficult to achieve. And I wondered why some of my colleagues are such surprisingly bad communicators, even within their own research communities, let alone on issues of public concern. Then I began to wonder what lucidity is, in a functional or operational sense. And then I began to suspect a deep connection with the way music works. Music is, after all, not only part of our culture but also part of our unconscious human nature.

I now like to understand the word 'lucidity' in a more general sense than usual. It's not only about what you can find in style manuals and in books on how to write, excellent and useful though many of them are. (Strunk and White[27] is a little gem.) It's also about deep connections not only with music but also with mathematics, pattern perception, biological evolution, and science in general. A common thread is what I call the *organic-change principle*.

The principle says that we're perceptually sensitive to, and have an unconscious interest in, patterns exhibiting 'organic change'. These are patterns in which some things change, continuously or by small amounts, while others stay the same. So an organically-changing pattern has *invariant elements*.

The walking dots animation is an example. The invariant elements include the number of dots, always twelve dots. Musical harmony is another.

Musical harmony is an interesting case because 'small amounts' is relevant not in one but in two different senses, as we'll see in Chapter 6. That leads to the idea of 'musical hyperspace'. An organic chord progression, or harmony change, can take us somewhere that's both nearby and far away. That's how some of the magic is done, in many genres of Western music. An octave leap is a large change in one sense, but small in the other, indeed so small that musicians use the same name for the two pitches. The invariant elements in an organic harmony change can be pitches or chord shapes.

Music makes use of organically-changing sound patterns not just in its harmony, but also in its melodic shapes and counterpoint and in the overall form, or architecture, of an entire piece of music. That's part of how it can grab our attention. Mathematics, too, contains organically-changing patterns. In mathematics, there are beautiful results about 'invariants' or 'conserved quantities', things that stay the same while other things change, often continuously through a vast space of possibilities. The great mathematician Emmy

Noether discovered a common origin for many such results, through a profound and original piece of mathematical thinking. Her discovery is called *Noether's Theorem* and is recognized today as a foundation-stone of theoretical physics.

Our perceptual sensitivity to organic change exists for strong biological reasons. One reason is the survival value of seeing the difference between living things and dead or inanimate things. To see a cat stalking a bird, or a flower opening, is to see organic change.

So I'd dare to describe our sensitivity to it as deeply instinctive. Many years ago, I saw a pet kitten suddenly die of some mysterious but acute disease — a sudden freezing into stillness. I'd never seen death before, but I remember feeling instantly sure of what had happened — ahead of conscious thought. And the ability to see the difference between living and dead has been shown to be well developed in human infants a few months old.

Notice how intimately involved, in all this, are ideas of a very *abstract* kind. The idea of some things changing while others stay invariant is itself highly abstract, as well as simple. It's abstract in the sense that vast numbers of possibilities are included. There are vast numbers — combinatorially large numbers — of organically-changing patterns, musical, mathematical, visual, and verbal. Here again, we're glimpsing the fact already hinted at, that the unconscious brain can handle many possibilities at once. We have an unconscious power of abstraction. That's almost the same as saying that we have unconscious mathematics. Mathematics is a precise means of handling many possibilities, many patterns, at once, in a self-consistent way, and of discovering surprising interconnections between them.

The walking dots animation shows that we have unconscious Euclidean geometry, the mathematics of angles and distances. There are combinatorially large numbers of arrangements of objects, at various angles and distances from one another. The roots of mathematics and logic lie far deeper, and are evolutionarily far more ancient, than they're usually thought to be. They're hundreds of millions of years more ancient than archaeology might suggest. In Chapter 6 I'll show that our unconscious mathematics includes, also, the mathematics underlying Noether's theorem, and I'll show how all this is related to Plato's world of perfect mathematical forms.

So I've been interested in lucidity, 'lucidity principles', and related matters in a sense that cuts deeper than, and goes far beyond, the niceties and pedantries of style manuals. But before anyone starts thinking that it's all about Plato and ivory-tower philosophy, let's remind ourselves of some harsh practical realities — as Plato would have done had he lived today. What I'm talking about is relevant not only to music, mathematics, thinking, and communication skills but also, for instance, to the ergonomic design of machinery, of software and user-friendly IT systems (information technology), of user interfaces in general, and of technological systems of any kind — not least the emerging artificial-intelligence systems, where the stakes are so incalculably high.

The organic-change principle — that we're perceptually sensitive to organically-changing patterns — shows why good practice in any of these endeavours involves not only variation but also invariant elements, i.e., repeated elements, just as music does. Good control-panel or website design might use, for instance, repeated shapes for control knobs or buttons. And in writing and speaking, one needn't be afraid of repetition if it forms the invariant element within an organically-changing word pattern. "If you are serious, then I'll be serious" is a clearer and stronger sentence than "If you are serious, then I'll be also." Loss of the invariant element "serious" weakens the sentence. Still weaker are versions like "If you are serious, then I'll be earnest." Such pointless or gratuitous variation in place of repetition is what H. W. Fowler ironically called "elegant" variation, an "incurable vice" of second-rate writers.[28] Its opposite can be called lucid repetition, as in "If you are serious, then I'll be serious." Lucid repetition is not the same as being repetitious. The pattern as a whole is changing, organically. It works the same way in every language I've looked at, including Chinese.[29]

Two more 'lucidity principles' are worth noting here. There's an explicitness principle — the need to be more explicit than you feel necessary — because, obviously, you're communicating with someone whose head isn't full of what your own head is full of. As the great mathematician J. E. Littlewood once put it,[30] "*Two trivialities omitted can add up to an impasse.*" Again, this applies to design in general, as well as to any form of writing or speaking that aims at lucidity. Quite often, all that's needed is to use a noun, perhaps repeated, rather than a pronoun. With a website button marked 'Cancel', it helps to say *what* it cancels. And then there's the more obvious coherent-ordering principle, the need to build context before new points are introduced. It applies not only to writing and speaking but also to the design of any sequential process on, for instance, a website or a ticket-vending machine.

One reason for attending to these principles is that human language is surprisingly weak on logic-checking, including checks for self-consistency.

That's one of the reasons why language is such a conceptual minefield — something that's long kept philosophers in business. And beyond everyday misunderstandings we have, of course, the workings of professional camouflage and deception, as in the ozone and other disinformation campaigns.

The logic-checking weakness shows up in the misnomers and self-contradictory terms encountered not only in everyday dealings but also — to my continual surprise — in the technical language used by my scientific and engineering colleagues. You'd think we should know better. You'd laugh if, echoing Spike Milligan, I said that someone has a hairy bald head. But consider, for example, the technical term 'solar constant'. It's a precisely defined measure of the mean solar power per unit area reaching the Earth. Well, the solar constant isn't constant. It's variable. The Sun's output is variable. We have a variable solar constant.

Also self-contradictory is the term 'slow manifold', a technical term used in my research field of atmospheric and oceanic fluid dynamics. Well, the slow manifold is a kind of hairy bald head. I'm really not kidding.[31]

In air-ticket booking systems there's a 'reference number' that isn't a number. In finance there's a term 'securitization' that means, among other things, making an investment less secure — yes, *less* secure — by camouflaging what it's based on. That particular misnomer contributed to the 2008 financial crash.

Being more explicit than you feel necessary helps you to navigate the minefield, when writing or speaking, and to correct inconsistencies. It clarifies your own thinking. Also helpful is to weed out gratuitous variations and replace them by lucid repetitions, maintaining the sometimes tricky discipline of calling the same thing by the same name, as in good control-panel design using repeated control-knob shapes.

It's even better to be cautious about choosing which shape, or which name or term, to use. You might even want to define a technical term carefully at its first occurrence, if only because meanings keep changing, even in science. 'I'll use the idea of whatsit in the sense of such-and-such, not to be confused with whatsit in the sense of so-and-so.' 'I'll denote the so-called solar constant by S, remembering that it's actually variable.' Another example is 'the' climate sensitivity. It has multiple meanings, as I'll explain in the

Postlude. In his 1959 Reith Lectures, Peter Medawar remarks on the "appalling confusion and waste of time" caused by the "innocent belief" that a single word should have a single, context-independent meaning.[32]

Another 'lucidity principle' — again applying to visual and technical design as well as to writing and speaking — is of course pruning, the elimination of anything superfluous. On your control panel, or web page, or ticket-vending machine, or in your software code and documentation, it's helpful to omit visual and verbal distractions. In writing and speaking, it's helpful to "omit needless words", as Ref. 27 puts it. You may have noticed lucidity principles in action in the meteoric rise of some businesses. Apple and Google were clear examples. There's a tendency to regard lucidity principles as trade secrets, or proprietary possessions. I recall some litigation by another fast-rising business, Amazon, in its early days, claiming proprietary ownership of 'omit needless clicks'.

Websites, ticket-vending machines, and other user interfaces that, by contrast, violate lucidity principles — making them 'user-unfriendly' — are still remarkably common, together with all those user-unfriendly technical manuals, and financial instruments securitized and unsecuritized. Needless complexity is mixed up with inexplicitness and gratuitous variation. The pre-Google search engines, with their cluttered screens and hidden, inexplicit search rules, were typical examples. But in case you think this is getting trivial, let me remind you of Three Mile Island Reactor TMI-2, and the nuclear accident for which it became well known in 1979.

You don't need to be a professional psychologist to appreciate the point. Before the nuclear accident, the control panels were like a gratuitously varied set of traffic lights in which stop is sometimes denoted by red and sometimes by green. Thus, at Three Mile Island, a particular colour on one control panel meant normal functioning, while the same colour on another panel meant 'malfunction, watch out'.[33] Well, the operators got confused and the nuclear accident happened, costing billions of dollars.

As I walk around Cambridge and other parts of the UK, I continually encounter the 'postmodernist traffic rules' followed by pedestrians here. Postmodernism says, in some versions at least, that 'anything goes'. So you keep left or keep right just as you fancy. All for the sake of interest and variety. How boring, how pedantic, to keep left all the time. Just like those boring traffic lights where red always means stop. To be fair, the UK *Highway Code* quite reasonably tells us to face oncoming traffic on narrow country lanes except, of course, on right-hand bends, and on unsegregated pedestrian-plus-cycle tracks where the *Code* does indeed say, implicitly, that anything goes. I've always felt a sense of relief when visiting the USA, where everyone keeps right most of the time.

And as for user-unfriendliness in technical manuals — well, the less said, the better. In some cases the authors seem to think that the user can cope with almost anything. They seem to be saying "Just read the manual and do what it says. Look, it says quite clearly that red means *stop* on one-way streets, *go* on two-way streets, and *caution* on right-hand bends. And of course it's the other way round on Sundays and public holidays, apart from Christmas, which is obviously an exception. What could be clearer? Just read it carefully and do exactly what it says," etc.

With complex systems like nuclear power plants, or large IT systems, or space-based observing systems — such as the wonderful Earth-observing systems created by my scientific colleagues — there's a combinatorially large number of ways for the system to go wrong even with good design, and even with communication failures kept to a minimum, in the technical manuals and elsewhere. I'm always amazed, and lost in admiration, when any of these systems work. I'm also amazed at how our governing politicians overlook the point again and again, it seems, when commissioning the large, centralized bureaucratic-plus-IT systems that they hope will work for them and save money.

A recent example was the UK's centralized test-and-trace system for the COVID-19 pandemic. The government thought it would be a "world-beating" system, as our then Prime Minister put it, that would quickly upstage, outperform, and replace the local test-and-trace systems already developed. But even a year after it was commissioned, the new centralized system remained largely ineffective, as the virus continued to spread and mutate. Far from saving money, by its failure the system must have added billions to the cost of the pandemic. Its design flaws included not only blindness to local information and knowhow but also, guess what, built-in user-unfriendliness — at human–human as well as at human–machine interfaces — generating widespread chaos and confusion.

What then is lucidity, in the sense I'm talking about — lucidity of writing, speaking, thinking, and design? Let me try to draw a few threads together. In the words of an earlier essay,[34] "Lucidity... exploits natural, biologically ancient perceptual sensitivities, such as the sensitivities to organic change and to coherent ordering, which reflect our instinctive, unconscious interest in the living world in which our ancestors survived. Lucidity exploits, for instance, the fact that organically-changing patterns contain invariant or repeated elements. Lucid writing and speaking are highly explicit, and where possible use the same word or phrase for the same thing" and similar word-patterns for similar things — lucid pattern-repetition as it might be called. "Context is built before new points are introduced..."

I also argued that "Lucidity is something that satisfies our unconscious, as well as our conscious, interest in coherence and self-consistency" — in things that make sense — and that it's about "making superficial patterns consistent with deeper patterns." It can be useful to think of our perceptual apparatus as a *multi-level* pattern recognition system, with many unconscious levels.

To summarize, then, five 'lucidity principles' seem especially useful in practice. They say that skilful communicators and designers pay close attention to the following:

1. organic change and lucid repetition,
2. explicitness,
3. coherent ordering,
4. self-consistency, and
5. pruning needless material.

These principles apply not only to writing and speaking but also, for instance, to website and smartphone design and to the safety systems of nuclear power plants, with stakes measured in billions of dollars.

Of course, a mastery of lucidity principles can also serve an interest in disinformation, with even higher stakes. Skilful liars are lucid. Such mastery was conspicuous in the ozone disinformation campaigns, and in more recent disinformation campaigns including those using weaponized postmodernism to spread confusion and 'alternative facts'.[35] Such mastery is, and always was, conspicuous in the dichotomizing speeches of demagogues, pressing our binary buttons as in 'You're either with us or against us'. That's another case of making superficial patterns consistent with deeper patterns — deeper patterns of an unpleasant and dangerous kind.

Enough of that! What of my other question? What *is* this subtle and elusive thing we call understanding, or insight? What does it mean to think clearly about a problem?

Of course there are many answers, depending on one's purpose and viewpoint. I'll focus on scientific understanding.

What I've always found in my own research, and have always tried to suggest to my students, is that developing an in-depth scientific understanding of something — understanding in detail how it works — requires looking at it, and testing it, from as many different viewpoints as possible. That's an important part of the creativity that goes into good science. And it puts a premium on good communication, including the ability to listen, actively, to someone offering a different viewpoint or focusing on a different aspect. Another part is to maintain a healthy scepticism, while respecting the evidence. And because it respects the evidence, such creativity is to be sharply distinguished from the postmodernist 'anything goes'.

For instance, the multi-timescale fluid dynamics I've worked on professionally is far too complex to be understandable at all from a single viewpoint, such as the viewpoint provided by a particular set of mathematical equations. One needs a multi-modal approach with equations, words, pictures, and feelings all working together, as far as possible, to form a self-consistent whole with experiments and observations. And the equations themselves take different forms embodying different viewpoints, with technical names such as 'variational', 'Eulerian', 'Lagrangian', and so on. They're mathematically equivalent but, as the great physicist Richard Feynman used to say, "psychologically very different". I'll give an example in Chapter 6. Bringing in words, in a lucid way, is an important part of the whole but needs to be related to, and made consistent with, equations, pictures, and feelings.

Such multi-modal thinking and healthy scepticism have been the only ways I've known of escaping from the mindsets and unconscious assumptions that tend to entrap us, and of avoiding false dichotomies in particular. The history of science shows that escaping from unconscious mindsets has always been a key part of progress, as already remarked,[5] including what Thomas Kuhn famously called paradigm shifts. And an important aid to cultivating a multi-modal view of any scientific problem is the habit of performing what Albert Einstein called thought-experiments, and mentally viewing *those* from as many angles as possible.

Einstein certainly talked about feeling things, in one's imagination — forces, motion, colliding particles, light waves — and was always doing thought-experiments, mental what-if experiments if you prefer. The same thread runs through the testimonies of Richard Feynman and of other great scientists, such as Peter Medawar and Jacques Monod. It all goes back to juvenile play — that deadly serious rehearsal for real life — curious young animals and children pushing and pulling things (and people!) to see, and feel, how they work.

In my own research community I've often noticed colleagues having futile arguments about 'the' cause of some phenomenon. "It's driven by such-and-such", says one. "No, it's driven by so-and-so", says another. Sometimes the argument gets quite acrimonious. Often, though, they're at cross-purposes because — perhaps unconsciously — they have different thought-experiments in mind.

Notice how the verb 'to drive' illustrates what I mean by language as a conceptual minefield. 'Drive' sounds incisive and clear-cut, but is nonetheless dangerously ambiguous. I sometimes think that our computers should make it flash red for danger, as soon as it's typed, along with some other dangerously ambiguous words such as the pronoun 'this'.

'To drive' can mean 'to control', as when driving a car, or controlling an audio amplifier via its input signal. But 'to drive' can also mean 'to supply the energy needed', via the fuel tank or the amplifier's power supply. Well, there are two quite different thought-experiments here, on the amplifier let's say. One is to change the input signal. The other is to switch the power off. A viewpoint focused on the power supply alone misses crucial aspects of the problem.

You may laugh, but there's been a mindset in my research community that has, or used to have, precisely such a focus. It said that *the* way to understand our atmosphere and oceans is through their intricate 'energy budgets', disregarding questions of what they're sensitive to. Yes, energy budgets are interesting and important, but no, they're not the Answer to Everything. Energy budgets focus attention on the power supply, making an input signal look unimportant just because it's small.

Instead of the verb 'to drive' it's often helpful, I think, to use the verb 'to mediate', as in the biological literature where it usually points to an important *part* of some mechanism.

The topic of mindsets and unconscious assumptions has been illuminated not only through the work of Kahneman and Tversky[6] but also through, for instance, that of Iain McGilchrist[36] and Vilayanur Ramachandran.[37] They bring in the workings of the brain's left and right hemispheres. That's a point to which I'll return in Chapter 4. In brief, the right hemisphere typically takes a holistic view of things and is more open to the unexpected, while the left hemisphere specializes in dissecting fine detail and is more prone to mindsets, including their unconscious aspects. The sort of scientific understanding I'm talking about — in-depth, multi-modal, multi-level understanding — seems to involve an intricate collaboration between the two hemispheres, with each playing to its own very different strengths.

Conversely, if that collaboration is disrupted by brain damage, extreme forms of mindset can result. Clinical neurologists are familiar with a delusional mindset called anosognosia. Damage to the right hemisphere paralyses, for instance, a patient's left arm, yet the patient vehemently denies that the arm is paralyzed, and will make all sorts of excuses as to why he or she doesn't fancy moving it when asked.

Back in the 1920s, the great physicist Max Born was immersed in the mind-blowing experience of developing quantum theory. Born later remarked that engagement with science and its healthy scepticism can give us an escape route from mindsets and unconscious assumptions. With the more dangerous kinds of zealotry or fundamentalism in mind, he said[38]

> "I believe that ideas such as absolute certitude, absolute exactness, final truth, etc., are figments of the imagination which should not be admissible in any field of science... This loosening of thinking *[Lockerung des Denkens]* seems to me to be the greatest blessing which modern science has given to us. For the belief in a single truth and in being the possessor thereof is the root cause of all evil in the world."

Further wisdom on these topics can be found in, for instance, the classic study of fundamentalist cults by Flo Conway and Jim Siegelman.[39] It echoes religious wars over the centuries. Time will tell, perhaps, how the dangers from the fundamentalist religions compare with those from the fundamentalist atheisms. Among today's fundamentalist atheisms we have not only scientific fundamentalism, saying that Science Is the Answer to Everything and Religion Must Be Destroyed — provoking a needless backlash against science, sometimes violent — but also, for instance, atheist versions of what economists now call market fundamentalism.[2, 25]

Market fundamentalism is arguably the most dangerous of all because of its financial and political power, still remarkably strong in today's world. I don't mean Adam Smith's reasonable idea that market forces and profits are *useful*, in symbiosis with the division of labour and good regulation.[25] Smith was clear about the need for regulation, written or unwritten.[40] I don't mean the business entrepreneurship that can provide us with useful goods and services. By market fundamentalism I mean the hypercredulous belief, the taking-for-granted, the simplistic and indeed incoherent mindset that market forces are by themselves the Answer to Everything, when based solely on 'deregulation' and the maximization of individual profit — regardless of evidence like the 2008 financial crash. Some adherents consider their beliefs 'scientifically' justified through the idea, which they wrongly attribute to Darwin, that competition between individuals is all that matters.[1] That last idea isn't, I should add, exclusive to the so-called political right.[2]

Understanding market fundamentalism is important because of its tendency to promote not only financial but also social instability, not least through gross economic inequality.[2] And the financial power of market fundamentalism makes it one of the greatest threats to good science, and indeed to rational problem-solving of any kind because, for a true believer, individual profit is paramount, taking precedence over respect for evidence — evidence about financial and social stability, or mental health, or pandemic viruses, or biodiversity, or the ozone hole or climate or anything else. The point is underlined by the investigations in Refs. 24 and 41.

Common to all forms of fundamentalism, or puritanism, or extremism is that, besides ignoring or cherry-picking evidence, they forbid the loosening of thinking that allows freedom to view things from more than one angle. Only one viewpoint is permitted, for otherwise you are 'impure'. You're commanded to have tunnel vision. The 2008 financial crash seems to have made only a small dent in market fundamentalism, so far, though perhaps reducing the numbers of its adherents. Perhaps the COVID-19 pandemic will make a bigger dent. It's still too early to say. And what's called 'science versus religion' is not, it seems to me, about scientific insight versus religious, or spiritual, insight. Rather, it's about scientific fundamentalism versus religious fundamentalism, which of course are irreconcilable.

Such futile dichotomizations cry out for more loosening of thinking. How can such loosening work? As Ramachandran or McGilchrist might say, it's almost as if the right brain hemisphere nudges the left with a wordless message to the effect that '*You* might be sure, but *I* smell a rat: could you, just possibly, be missing something?'

It's well known that in 1983 a Russian officer, Stanislav Petrov, saved us from likely nuclear war. At great personal cost, he disobeyed standing orders when a malfunctioning weapons system said 'nuclear attack imminent'. He smelt a rat and we had a narrow escape. We probably owe it to Petrov's right hemisphere. There have been other such escapes.

3

Mindsets, evolution, and language

Let's fast-rewind to a few million years ago, and further consider our ancestors' evolution. Where did we, our insights, and our mindsets come from? And how on Earth did we acquire our language ability — that vast conceptual minefield — so powerful, so versatile, yet so weak on logic-checking? These questions are more than just tantalizing. Clearly, they're germane to past and present conflicts, and to future risks including existential risks.

The first obstacle to understanding is what I'll dare — following a suggestion by John Sulston — to call *simplistic evolutionary theory*. The theory is still firmly entrenched in popular culture, with labels like 'Darwinian struggle'. Many biologists would now agree with John that the theory is no more than a caricature. But it's a remarkably persistent caricature. It's still hugely influential. It includes the idea that competition between individuals is all that matters.

More precisely, simplistic evolutionary theory says that evolution has just three aspects. First, the structure of an organism is governed entirely by its genome, acting as a deterministic 'blueprint' made of all-powerful 'selfish genes'. Second, contrary to what Charles Darwin thought,[18] natural selection is the *only* significant evolutionary force. And third, natural selection works through 'survival of the fittest', conceived of solely in terms of a struggle between individuals.

Survival of the fittest would be a reasonable proposition were it not that an oversimplified notion of fitness is used. Not only is fitness presumed to apply solely to individual organisms, but it's also presumed to mean nothing more than the individual's ability to pass on its genes. Admittedly this purely competitive, purely individualistic view does explain much of what happens in our planet's astonishing biosphere. But it also misses many crucial points. It's not the evolutionary Answer to Everything.

There's a slightly more sophisticated view called 'inclusive fitness' or 'kin selection', which replaces individuals by families whose members share enough genes to count as closely related. But it misses the same points.

For one thing, as Darwin recognized, our species and many other social species, such as baboons, could not have survived without cooperation within large groups. Without such cooperation, alongside competition, our ground-dwelling ancestors would have been easy meals for the large, swift predators all around them, including the big cats — gobbled up in no time at all! Cooperation restricted to a few closely related individuals would not have been enough to survive those dangers. And Darwin gives clear examples in which cooperation within large non-human groups is, in fact, observed to take place, as with the geladas and the hamadryas baboons of Ethiopia.[1]

Even bacteria cooperate. That's well known. One way they do it is by sharing small packages of genetic information called plasmids or DNA cassettes. A plasmid might, for instance, contain information on how to survive antibiotic attack. Don't get me wrong. I'm not saying that bacteria think like us, or like baboons or dolphins or other social mammals, or like social insects. And I'm not saying that bacteria never compete. They often do. But for instance it's a hard fact — a practical certainty, and now an urgent problem in medicine — that large groups of individual bacteria cooperate among themselves to develop resistance to antibiotics. For the bacteria, such resistance is highly adaptive, and strongly selected for. Yes, selective pressures are at work, but at group level as well as at individual and kin levels, and at cellular and molecular levels,[18] in heterogeneous populations living in heterogeneous, and ever-changing, ecological environments.

So it's plain that natural selection operates at many levels within the biosphere, and that cooperation is widespread alongside competition. Indeed, the word 'symbiosis' in its standard meaning denotes a variety of intimate, and well-studied, forms of cooperation not between individuals of one species but between those of entirely different species. And different species of bacteria share plasmids.[42] The trouble is the sheer complexity of it all — again a matter of combinatorial largeness as well as of population heterogeneity, and of the complexities of mutual fitness in and around various ecological niches. We're far from having comprehensive mathematical models of how it all works.

On the other hand, though, the models have made great progress in recent decades, helped by increasing computer power. There have been significant advances at molecular level.[18] They've added to the accumulated evidence for what's now called multi-level natural selection or, for brevity, multi-level selection.[43-55] The evidence comes not only from better models at various levels but also, for instance, from laboratory experiments with heterogeneous populations of real organisms, directly demonstrating group-level selection.[43]

The persistence of simplistic evolutionary theory, oblivious to all these considerations, seems to be bound up with a particular pair of mindsets. The first says that the genes' eye view — or, more fundamentally, the replicators' or DNA's eye view — gives us the *only useful angle* from which to view evolution. The second reiterates that selective pressures operate at one level only, that of individual organisms. The first mindset misses the value of viewing a problem from more than one angle. The second misses most of the real complexity. And that complexity includes not only group-level selective pressures as demonstrated in the laboratory,[43] but also the group-level selective pressures on our ancestors noted by biologists such as Jacques Monod[3] and by palaeoanthropologists such as Phillip Tobias,[4] Robin Dunbar,[46] and Matt Rossano[48] to name but a few.

Both mindsets seem to have come from mathematical models that are grossly oversimplified by today's standards, valuable though they were in their time. They are the old population-genetics models that were first formulated in the early twentieth century[44] and then further developed in the 1960s and 1970s. For the sake of mathematical simplicity and solvability those models exclude, by assumption, all the aforementioned complexities as well as multi-timescale processes and, in particular, realistic group-level selection scenarios.[43, 52, 54] And the hypothetical 'genes' in those models are themselves grossly oversimplified. They correspond to the popular idea of a gene 'for' this or that trait — nothing like actual genes, the protein-coding sequences within the genomic DNA. Very many actual genes are involved, usually, in the development of a recognizable trait, along with non-coding parts of the DNA, the associated regulatory networks, and the environmental circumstances.[3, 9, 18]

The mindset against group-level selection, built into the old models and still strongly held in some circles,[12, 56, 57] seems in part to have been a reaction against the sloppiness of some old arguments in favour of such selection, for instance ignoring the complex game-theoretic aspects such as multi-agent 'reciprocal altruism', and conflating altruism as a conscious anthropomorphic sentiment 'for the good of the group' with altruism as actual behaviour, including its deeply unconscious aspects.[1, 2, 10, 11, 43] Among the arguments against group-level selection one can find phrases such as 'mathematical *proofs* from population genetics' (my italics).[56, 57] But such 'proofs' rely on the equations of the old models. More recent population-genetics models do, by contrast, incorporate realistic group-level selection, and identify examples in which it's significant.[52, 54]

On the multi-timescale aspects, there's some fascinating history. Despite those aspects having been recognized back in 1971, both by Monod[3] and by Tobias,[4] and noted again in some recent reviews,[47, 51, 53] it's hard to find mention of them elsewhere in the literature on genome–culture co-evolution. They may have been lost in the turmoil of other disputes about biological evolution. Some of those disputes are described in the remarkable book by Christopher Wills,[44] and more of them in the vast sociological study by Ullica Segerstråle.[58]

Two of the most acrimonious disputes, going back to the 1970s, are germane to our discussion. The first surrounded a famous attempt by Edward O. Wilson and Charles J. Lumsden to model genome–culture co-evolution *without* multiple timescales. Instead, a mysterious 'multiplier effect' was proposed, which drastically shortens genomic timescales. Then, in the model, genome and culture can co-evolve in a simple, quasi-synchronous way.

Such model scenarios are now recognized as unrealistic. They're inconsistent with the genomic and linguistic evidence, as we'll see.

The second dispute was about adaptive mutations to genomes — mutations that give an organism an immediate advantage — versus neutral mutations, which have no immediate effect and are therefore invisible to natural selection. Dichotomization kicked in straight away, with protagonists assuming one thing to the exclusion of the other.[44] But we now know that both are important. For instance, the recent work described in the book by Andreas Wagner[18] shows in detail, at molecular level, why both kinds of mutation are practically speaking inevitable, and how neutral mutations generate genetic diversity. Genetic diversity is required to let natural selection work in ever-changing environments. What started as neutral can *become* adaptive. In Chapter 4, I'll suggest that brain-hemisphere differentiation is a likely example. So as Darwin said, presciently, natural selection is not the only significant evolutionary force.[18] None of this would need saying yet again, were it not for the extraordinary persistence of simplistic evolutionary theory.

On the palaeoanthropological side, there's been a tendency to see the technology of stone tools as the only important aspect of 'culture', giving an impression that culture stood still for a million years or more, just because the stone tools didn't change very much. These were the Acheulean bifacial hand axes. They were made by striking flakes off brittle stones such as obsidian or flint, calling for highly developed skills. But it's worth stressing again, I think, that tool shapes and an absence of beads and bracelets, and of other such archaeological durables, is no evidence at all for an absence, or a standing-still, of culture and of language or proto-language whether gestural, or vocal, or both. That's because, as said before, language and culture can be mediated purely by sound waves and light waves, leaving no archaeological trace.

Arguably, some of the selective pressures for language development could have included pressures to improve the teaching of toolmaking skills.[59, 60]

And what of that other feature of simplistic evolutionary theory, the genome seen as a blueprint dictating how an organism develops — as in *Jurassic Park*? Well, that's been superseded, and in many ways discredited, by detailed studies of the workings of the genome that recognize the systems-biological aspects.[3, 9, 18, 55, 57] Multiple layers of molecular-biological complexity — of regulatory networks — overlie the genome. Causal arrows point downward as well as upward. Far from dictating everything, genes are switched on and off or otherwise regulated according to need. That's why brain cells differ from liver cells. The genomic DNA is highly influential but not dictatorial.

It would be more accurate to say that the DNA provides something like a toolkit — a toolkit for building and repairing a particular organism. Or better still, the DNA and its associated regulatory networks, or biomolecular circuits, provide a set of self-organizing or self-assembling building blocks[55, 57] available for use according to developmental and environmental need. Echoing the language of complexity theory touched on in Chapter 1, I'll call these self-assembling building blocks *genetically enabled automata*.

* * *

How, then, might group-level selection have worked for our ancestors, over recent millions of years? I want to avoid good-of-the-group sloppiness, but I think it reasonable, following Monod,[3] Tobias,[4] Dunbar,[46] and Rossano[48] — and, for instance, an illuminating survey by Leslie Aiello[61] — to suggest that not only language but also the gradually developing ability to create what eventually became complex music, dance, poetry, rhetoric, and storytelling, held in the memories of gifted individuals and in a group's collective, cultural memory — and leaving no archaeological trace — would have produced selective pressures for further brain development and for generations of individuals still more gifted.

Not only the best toolmakers, hunters, fighters, and tacticians but also, in due course, the best and most empathetic singers and storytellers — or more accurately, perhaps, *singer-storytellers* — would have had the best mating opportunities. Intimately part of all this would have been the kind of social intelligence we call 'theory of mind' — the ability to feel what others are thinking and, in due course, what others think I'm thinking, and so on, alongside new abilities to imagine future events,[59, 60] and fictitious events.[44, 48, 62] Those developments would have increased group size and solidarity and the power of a group to compete against other groups for resources, as well as to ward off predators and, in due course, to become the top predators themselves — as groups, though, and not as individuals.[48]

We need not argue about the timescales and the precise stages of development except to say that their beginnings must have gone back to times far more ancient than the past several tens of millennia that we call the Upper Palaeolithic, with its variety of archaeological durables such as beads, bracelets, statuettes, bone flutes, cave paintings and other beautiful art objects signalling what's been been called the 'cognitive revolution'.[62]

Such archaeological evidence has sometimes been taken to mark the beginning of language.[12, 13] However, much longer timescales for language development are dictated by the rates at which genomic changes occur, as evidenced by molecular-genetic clocks.[44]

In particular, there had to be enough time to allow the self-assembling building blocks for language to grow within genetic memory — the genetically enabled automata for language — from rudimentary beginnings or proto-language as echoed, perhaps, in the speech and gestural signing of a two-year-old human today and perhaps contributing, as already suggested, to the development and teaching of early toolmaking.[59, 60] But, in any case, the existence of such automata has been conclusively demonstrated by recent events in Nicaragua, of which we'll be reminded shortly. The proposition that language was a purely cultural invention and began just before the Upper Palaeolithic[12, 13] is now, therefore, completely untenable.

So the earliest stages of all this, perhaps a million years ago or more, would have seen language barriers beginning to be significant — gradually sharpening the separation of one group from another. The groups, regarded as evolutionary 'survival vehicles', would have developed increasingly tight outer boundaries. Such boundaries would have enhanced the efficiency of those vehicles as carriers of replicators into future generations within each group. And, contrary to what is sometimes argued,[12] the replicators would have been genomic as well as cultural. Little by little, this tendency to channel both kinds of replicator within groups down the generations must have increasingly strengthened the feedback — the multi-timescale, two-way coupling — between genomes and cultures. And the same channelling would have intensified the selective pressures exerted at group-versus-group level, just as Monod[3] and others have argued. Genomic group selection and cultural group selection, as they're called, would have been slow and fast mechanisms within a single ongoing process whose end result was the runaway brain evolution, as Wills[44] called it, corresponding to the left-hand extremity of Figure 2:

Human and pre-human brain sizes over the past 3,500 millennia

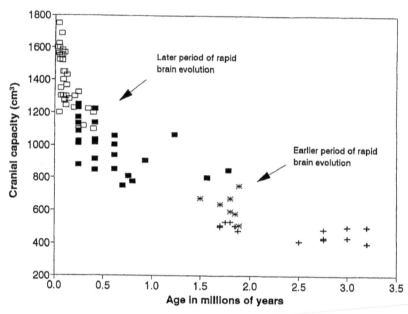

Figure 2: Skull capacities of some of our ancestors and their close relatives as found in the fossil record, from Ref. 61. Time runs from right to left. The Acheulean toolmaking tradition began around the time of early *Homo erectus*, the earliest black squares. Pluses denote our bipedal australopithecine ancestors or near-ancestors — Monod[3] calls them australanthropes. Asterisks denote early *Homo* (*H. habilis*, *H. rudolfensis*), black squares, *H. erectus* including *H. ergaster*, and open squares archaic and modern *H. sapiens*. The same overall pattern is seen in more recent data compilations.[63] (This figure © The British Academy 1996, reproduced by permission from the *Proceedings of the British Academy*.)

The final, spectacular acceleration in brain size seen on the left of Figure 2 must have been associated with cultures and languages of ever-increasing complexity, setting the stage around 100 millennia ago for the so-called cognitive revolution of the Upper Palaeolithic and the explosion of imaginative artwork that accompanied it, increasingly visible in the archaeological record. Although that explosion cannot have marked the beginning of language, as such, it could well have marked important new developments in the *use* of language and in the associated cognitive abilities including 'theory of mind'. Paramount among those new abilities would have been mythmaking — storytelling about fictitious characters and fictitious, imagined worlds — opening up radical new possibilities for versatile cooperative behaviour and for building larger and more powerful groups, and group identities, than ever before.[44, 48, 62]

Looking back over the sequence of events echoed in Figure 2, Aiello[61] pictures them as having come from a long-drawn-out "evolutionary arms race" between competing groups or tribes, long predating the Upper Palaeolithic. Such an arms race would have put an early premium on increasing group size, as well as on language-assisted solidarity and coordination within a group.[46] To be sure, any inter-group or inter-tribal warfare that took place could have diluted the within-group channelling of genomic information through, for instance, the enslavement of enemy females. But then again, that's a further evolutionary advantage. It's a one-way gene flow *into* the biggest, strongest, smartest groups or tribes, adding to their genetic diversity and adaptive potential as soon as it spills across internal caste boundaries and slave boundaries. Indeed, it's now known that our own genetic diversity contains DNA from distinct strands in our species' recent ancestry, including the Neanderthals and the Denisovans.[63]

All this is, of course, invisible to simplistic evolutionary theory. More realistic views of evolution can be found in the recent research literature,[43–55] sharpening our understanding of the multifarious dynamical mechanisms and feedbacks that can come into play. Some of those mechanisms have been discussed under headings such as 'evo-devo' (evolutionary developmental biology) and the 'extended evolutionary synthesis'. They include new mechanisms of 'epigenetic' heritability that operate outside the DNA sequences themselves[53] — all of which says that genetic memory and genetically enabled automata are even more versatile and flexible than previously thought.

Some researchers today go so far as to question the very use of the phrase 'genetic memory'. However, I think it remains useful as a pointer to the important contribution from the DNA-mediated information flow, alongside many other forms of information flow into future generations including flows via culture and via physical 'niche construction', as with earthworms, beavers, and countless other species. And as for the idea of genetically enabled automata, it seems to me so important — not least as an antidote to the simplistic genetic-blueprint idea — that I propose to use it without further apology.

The Nicaraguan evidence shows that human children are born with an innate potential for language, as we'll see shortly. What was observed is impossible to explain without genetically enabled automata. Please note that this is quite different from saying that language is innately 'hard-wired' or 'blueprinted'. The automata — the self-assembling building blocks — are not the same thing as the assembled product, assembled of course in a particular environment, physical and cultural, from versatile and flexible components.

Recognition of the distinction between building blocks and assembled product might even, I dare to hope, get us away from the dichotomized quarrels about 'all in the genes' versus 'all down to culture'. Yes, language is in the genes and regulatory DNA *and* culturally constructed, where of course we must understand the 'construction' as being largely unconscious, as great artists, great writers, and great scientists have always recognized — consciously or unconsciously. And there's no conflict with the many painstaking studies of comparative linguistics, showing the likely pathways and relatively short timescales for the cultural ancestry of today's languages.[12] Particular language groups, such as Indo-European, are one thing, while the innate potential for language is another.

But first, what about those multi-timescale aspects? How on Earth *can* genome, language, and culture co-evolve, and interact dynamically, when their timescales are so very, very different? And how, above all, can the latest cultural whim or flash in the pan influence so slow a process as genomic evolution? Isn't the comparison with air pressure far and away too simplistic?

Well, as already said there are countless other examples of multi-timescale processes in the natural world. Many are far more complex than air pressure. They can be even more extreme in their range of timescales. The Antarctic ozone hole is one such example that I happen to know about in detail. One might equally well ask, how can the very fast and very slow processes involved in the ozone hole have any significant interplay? How can the seemingly random turbulence that makes us fasten our seat belts have any role in a stratospheric phenomenon on a spatial scale bigger than the Antarctic continent and dependent on global-scale air motion, over a range of timescales out to very many decades?

As I was forced to recognize in my own research there *is*, however, a significant and systematic interplay between atmospheric turbulence and the ozone hole. That interplay is now well understood. Among other things, it involves a sort of fluid-dynamical jigsaw puzzle made up of breaking waves and turbulence. Despite differences of detail, and greater complexity, it's a bit like what happens in the surf zone near an ocean beach. There, tiny, fleeting eddies within the foamy turbulent wavecrests not only modify, but are also shaped by, the wave dynamics in an intricate interplay that, in turn, generates and interacts with mean currents, including dangerous rip currents, and with sand and sediment transport over far, far longer timescales.

The ozone hole involves two very different kinds of turbulence. One is the familiar small-scale, seat-belt-fastening turbulence, on timescales of seconds. The other is a much slower phenomenon on a much larger scale, the scale of a weather map, involving a chaotic interplay between jet streams, cyclones, and anticyclones. Several kinds of waves are involved, including large jet stream meanders of the kind familiar from weather forecasts. These are called Rossby waves. In the stratosphere, such meanders can develop into the "world's largest breaking waves". And interwoven with all that fluid-dynamical complexity we have regions with different chemical compositions, and an interplay between the transport of chemicals from place to place, on the one hand, and a large set of fast and slow chemical reactions on the other. The chemistry interacts with solar and terrestrial radiation, from the ultraviolet to the infrared, over a vast range of timescales from thousand-trillionths of a second as photons hit air molecules out to days, months, years, and decades as chemicals are moved around by the air motion. The key point about all this, though, is that what look like chaotic, flash-in-the-pan, fleeting, and almost random processes on the shorter timescales have systematic mean effects over far, far longer timescales. The observed ozone hole is just one of those mean effects.

In a similar way, then, our latest cultural whims and catch-phrases may seem capricious and fleeting and sometimes almost random — almost a 'cultural turbulence' — while nevertheless exerting systematic, long-term selective pressures that, as already suggested, favour the talents of gifted and versatile individuals who can influence and reshape the arts of communication, storytelling, imagery, politics, music, dance, and comedy, with storytelling emerging as the most basic and powerful of those arts. The feeling that it's all down to culture surely reflects the near-impossibility of imagining the vast overall timespans, out to millions of years, over which the genetically enabled automata that mediate language and culture must have evolved under those turbulent selective pressures, all the way from rudimentary beginnings.

And, as already said, the existence of genetically enabled automata for language — self-assembling building blocks[55, 57] for language — has been verified, conclusively, by recent events in Nicaragua.

Starting in the late 1970s, Nicaragua saw the emergence of a new Deaf community and a fully fledged, syntactically powerful new sign language, Nicaraguan Sign Language (NSL). Beyond superficial borrowings, NSL is considered by sign-language experts to be entirely distinct from any pre-existing sign language, such as American Sign Language or British Sign Language. It's clear, moreover, that NSL emerged from newly formed communities of deaf schoolchildren with essentially no external linguistic input. NSL could not have been taught by any adult. It must have self-assembled within the children's unconscious brains, in a process *enabled* by their genomes and externally stimulated by nothing more than the need to communicate with the other deaf children, in a newly established school environment.

Deaf people had no schools or substantial communities in Nicaragua before the late 1970s, a time of drastic political change. It was only then that dozens, then hundreds, of deaf children first came into social contact. This came about through a new educational programme, which established schools for the deaf. Today, full NSL fluency at native-speaker level, or rather native-signer level, is found in just one group of people. They are those, and only those, who were young children in the Deaf schools and communities that began in the late 1970s. That's a simple fact on the ground, as is the nonexistence of NSL in earlier times. It's therefore indisputable that NSL was created by the children.

The evidence[64] clearly shows that the youngest children aged about 7 or less played a crucial role in creating NSL, with no linguistic input from their Spanish-speaking surroundings. NSL has no particular resemblance to Spanish. A key aspect must have been a small child's instinct to impose syntactic function on whatever language is being acquired or created. After all, it's commonly observed that a small child acquiring English will say things like 'I eated the cake' instead of 'I ate the cake'. It's the syntactic irregularities that need to be taught by older people, not the syntactic function itself.

That last point was made long ago by Noam Chomsky among others. But the way it fits in with natural selection was unclear at the time. We didn't then have today's insights into multi-level natural selection, multi-timescale genome–culture feedback, and genetically enabled automata.

Regarding the detailed evidence from Nicaragua, the extensive account in Ref. 64 is a landmark. It describes careful and systematic studies using video and transcription techniques developed by sign-language experts. Those studies brought to light, for instance, what are called the pidgin and creole stages in the collective creation of NSL by, respectively, the older and the younger children, with full syntactic functionality arising at the creole stage only, coming from the children aged 7 or less. Ref. 65 gives a popular account of developments up to the early 1990s. More recent work[66] shows how the repertoire of syntactic functions in NSL has been filled out, and increasingly standardized, by successive generations of young signers.

* * *

And what of the changing climate that our ancestors had to cope with? Over the timespan of Figure 2, the climate system underwent increasingly large fluctuations some of which were very sudden, as will be illustrated shortly, drastically affecting our ancestors' food supplies and living conditions. In the later stages, which culminated in the runaway brain evolution and global-scale migration of our species, the increasing climate fluctuations would have been ramping up the pressure to develop tribal solidarity and versatility mediated by ever more elaborate mythologies, rituals, songs, and stories passed from generation to generation.

And what stories they must have been! Great sagas etched into a tribe's collective memory. It can hardly be accidental that the sagas known today tell of years of famine and years of plenty, of battles, of epic journeys, of great floods, and of terrifying deities that are both fickle benefactors and devouring monsters — just as the surrounding large predators must have appeared to our early ancestors, as those ancestors scavenged on leftover carcasses long before becoming top predators themselves.[67] And epic journeys and great floods must have been increasingly part of our ancestors' struggle to survive, as they migrated under the increasingly changeable climatic conditions.

Figure 3 is a palaeoclimatic record giving a coarse-grain overview of climate variability going back 800 millennia, roughly corresponding to the leftmost quarter of Figure 2. Again, time runs from right to left:

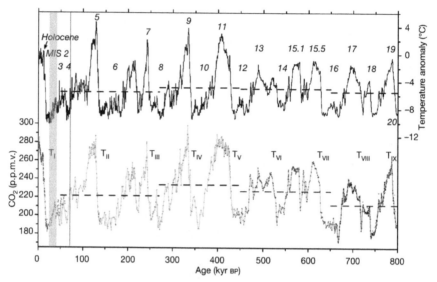

Figure 3: Antarctic ice-core data from Ref. 68 showing estimated temperature (upper graph) and measured atmospheric carbon dioxide (lower graph). Time, in millennia, runs from right to left up to just before the industrial era. The significance of the lower graph will be discussed in the *Postlude*. The upper graph estimates air temperature changes over Antarctica, indicative of worldwide changes. The temperature changes are estimated from the amount of deuterium (hydrogen-2 isotope) in the ice, which can be reliably measured and is temperature-sensitive because of fractionation effects as water evaporates and precipitates, and thereby redistributes itself between oceans, atmosphere, and ice sheets. (The 'MIS' numbers denote what palaeoclimatologists call 'marine isotope stages' whose signatures are recognized in many deep-ocean mud cores, and 'T' means 'termination' or 'deglaciation', the termination of a cold period.) The thin vertical line at around 70 millennia marks the time of the Lake Toba supervolcanic eruption. The shaded bar corresponds to the relatively short time interval covered in Figure 4 below. Reproduced by permission from Springer Nature Customer Service Centre GmbH: *Nature*,[68] © Springer Nature 2008.

The upper graph shows an estimate of temperature changes. The label 'Holocene' marks the relatively warm, stable climate of the past 10 millennia or so. The temperature changes are estimated from a reliable record in Antarctic ice (see Figure 3 caption) and are indicative of worldwide, not just Antarctic, temperature changes. There are questions of detail and precise magnitudes, but little doubt as to the *order of magnitude* of the estimated changes. The changes were huge, especially during the past four hundred millennia, with peak-to-peak excursions of the order of ten degrees Celsius or more. These excursions mark what are called the geologically recent glacial–interglacial cycles.

A good cross-check is that global mean sea-level changes were also huge, as temperatures went up and down and the great land-based ice sheets shrank and grew. For instance, sea levels rose by well over a hundred metres during the transition to the Holocene, and during the earlier transitions marked 5 and 9 in the upper graph. Land bridges, such as those between Siberia and Alaska and between England and Europe, disappeared under the waves. These are huge sea-level changes by any standard. They're certainly huge by comparison with sea-level changes in recent historical times.

The evidence on sea-level change is discussed in the *Postlude* below. Also discussed there is the significance of the lower graph, which shows concentrations of carbon dioxide in the atmosphere. Carbon dioxide as a gas is extremely stable chemically, allowing it to be reliably measured from the air trapped in the Antarctic ice.

When we zoom in to much shorter timescales, we see that some climate changes were not only severe but also abrupt, over time intervals comparable to, or even shorter than, an individual human lifetime. We know this thanks to a large body of work on the records in ice cores and oceanic mud cores, and in many other palaeoclimatic records.[69–71] The sheer skill and meticulous labour of fine-sampling, assaying, and carefully decoding such material to improve the time resolution, and to cross-check the interpretation, is a remarkable story of high scientific endeavour.

Not only were there occasional nuclear-winter-like events from volcanic eruptions, including the Lake Toba supervolcanic eruption around 70 millennia ago, at the thin vertical line in Figure 3 — a far more massive eruption than any in recorded history — but there was large-amplitude internal variability within the climate system itself. Even without volcanoes the system has so-called chaotic dynamics, with scope for rapid changes in, for instance, sea-ice cover and in the meanderings of the great atmospheric jet streams and their oceanic cousins, such as the Gulf Stream and the Kuroshio and Agulhas currents.

The chaotic dynamics produced sudden and drastic climate change over time intervals that were sometimes as short as a few years — practically instantaneous by geological, palaeoclimatic, and evolutionary standards. Such events are called 'tipping points' of the dynamics. They'll be further discussed in the *Postlude*. Much of the drastic variability now known — in its finest detail for the last fifty millennia or so — took the form of complex and irregular 'Dansgaard–Oeschger cycles'. They're conspicuous in the palaeoclimatic records over much of the northern hemisphere. See for instance Figure 4:

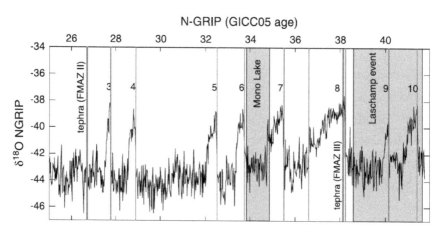

Figure 4: Greenland ice-core data from Dokken et al.,[71] for the time interval corresponding to the shaded bar in Figure 3. Time in millennia runs from right

Caption continues on facing page

to left. The graph shows variations in the amount of the oxygen-18 isotope in the ice, from which temperature changes can be estimated in much the same way as in Figure 3. The abrupt warmings marked by the thin vertical lines are mostly of the order of 10°C or more and were accompanied by the sudden disappearance of Nordic Sea ice, recorded in an ocean sediment core. The thicker vertical lines show timing checks from layers of tephra or volcanic debris. The shaded areas refer to geomagnetic excursions. Ref. 71 is open-access and available at https://doi.org/10.1002/palo.20042, copyright © American Geophysical Union, reproduced by kind permission.

In the figure, N-GRIP refers to a location in North Greenland, while GICC05 refers to the dating method. The graph is a record from Greenland ice with enough time resolution to show some of the Dansgaard–Oeschger cycles, those conventionally numbered from 3 to 10. The graph estimates air temperature changes over Greenland (see Figure 4 caption). The thin vertical lines mark the times of major warming events in the North Atlantic region, which by convention define the end of one cycle and the start of the next. Those 'Dansgaard–Oeschger warmings' were huge, typically of the order of ten degrees Celsius, in round numbers, as well as very abrupt. Indeed, they were far more abrupt than the graph can show. In some cases, they took place over a few decades only, stepwise with individual steps each taking a few years.[69, 71] Between the major warming events we see an incessant variability at more modest amplitudes — more like a few degrees Celsius — nevertheless more than enough to have affected our ancestors' food supplies and living conditions.

To survive all this, our ancestors must have had not just tribal solidarity but also, at least in times of crisis, strong leaders and willing followers. Hypercredulity and weak logic-checking must have had a role — selected for as genome and culture co-evolved and as language became more sophisticated, and more fluent and imaginative with stories of the natural and the supernatural.

How do you make leadership work? Do you make a reasoned case? Do you ask your followers to check your logic? Do you check it yourself? Of course not! You're a leader because, with your people starving, you've emerged as a charismatic visionary. You're divinely inspired. Your people love you. You *know* you're right, and it doesn't need checking. You have the Answer to Everything. "O my people, I've been shown the True Path that we must follow. Come with me! Let's make our tribe great again! Beyond those mountains, over that horizon, that's where we'll find our Promised Land. It is our destiny to find that Land and overcome all enemies because we, and only we, are the True Believers. Our stories are the only *true* stories." How else, in the incessantly fluctuating climate, I ask again, did our one species — our single human genome — spread all around the globe in less than a hundred millennia?

And what of dichotomization — that ever-present, ever-potent source of unconscious assumptions — assumptions that are so often wrong in today's world? Well, it's even more ancient. Hundreds of millions of years more ancient. Ever since the Cambrian, half a billion years ago, survival has teetered on the brink of edible or inedible, male or female, friend or foe, and fight or flight. In life-threatening emergencies, binary snap decisions were crucial. But with language and hypercredulity in place, the dichotomization instinct — rooted in the most ancient, the most primitive, the most reptilian parts of our brains — could grow into new forms. Not just friend or foe but also *We are right and they are wrong*. It's the Absolute Truth of our belief system versus the absolute falsehood of theirs, with nothing in between.

You might be tempted to dismiss the last point as a mere 'just so story', speculation unsupported by evidence. But I'd argue that there's plenty of evidence today. One clear line of evidence is the ease with which polarized conflicts are amplified, expanded, and intensified by the social media, despite the immense harm that they cause.

There's also the evidence documented in the book by evolutionary biologist David Sloan Wilson.[43] This carefully argued book includes case studies of fundamentalist or puritanical belief systems — religious in Chapter 6, and atheist in Chapter 7. For instance Ayn Rand, an atheist prophet revered for her preaching of market fundamentalism, held that selfishness is an absolute good and altruism an absolute evil. Personal greed at others' expense is the Answer to Everything, and the pinnacle of moral virtue. Wilson describes how a well-intentioned believer, Rand's disciple Alan Greenspan, was "dumbfounded" by the 2008 financial crash shortly after his long reign at the US Federal Reserve Bank. How could a system based on Rand's gospel fail so abysmally? By a supreme irony, Rand's gospel also says not only that 'We are right and they are wrong' but also that 'We are rational and they are irrational'. Any *logic-checking* that considers an alternative viewpoint is 'irrational', something to be dismissed out of hand.

It's the same for any other puritanical mindset. Only one viewpoint is permitted; and unless you dismiss the alternatives immediately, without stopping to think, you're *impure*, aren't you. You're a ditherer, a lily-livered moral weakling. You're in danger of sending a heretical 'mixed message' when it's all about Us Versus Them, Truth Versus Falsehood, Good Versus Evil. That's the force behind the so-called 'purity spirals' seen on social media, in which reasoned debates turn into shouting matches between increasingly polarized factions.

* * *

Yes, dichotomization makes us stupid. Luckily, however, dichotomization isn't all there is. We don't have to see the world this way. It's a trap we needn't fall into. Outside life-threatening emergencies, and far more than your average reptile, we do have the ability to stop and think. We do have the ability to see that while some issues are dichotomous, many others are not. We do have the ability to see — oh shock horror — that there might be merit in different viewpoints, helping us to solve real problems. And our hopes of a civilized future will depend, it seems to me, upon strengthening those thinking abilities. In the first edition of this book I ventured to hope that, surprising though it may seem, the social-media giants might, in their own self-interest, help with such strengthening.

For many years now the social-media giants, beginning with Facebook, have amassed vast wealth and supranational power by using the artificial intelligences, the robots, that they've built and trained within their secret 'large hadron collider of experimental psychology' as I called it — secret experimentation on billions of human subjects.[23] But even today those robots are not yet very intelligent! In some ways *they're* downright stupid. Why do I say that? Because, as currently set up, they pose an existential threat to their own survival along with that of the social media themselves, in their present-day entrepreneurial form.

Part of the threat comes from the way the robots have been trained to exploit the dichotomization instinct, alongside other primitive instincts such as fear, anger, and hatred, to make things go viral — reaching vast audiences and raking in huge profits from advertising revenue. Like or dislike, friend or unfriend, follow or don't follow, share or don't share, include or exclude, and so on, are not only addictive, attention-grabbing data-gathering devices but also, in their own way, examples of what I called 'perilous binary buttons'. Press or don't press, and don't stop to think. In this and in other ways of shutting off thinking the social-media robots have become, among other things, powerful disinformation-for-profit and hatred-for-profit machines. On a massive and unprecedented scale they've amplified fear, anger, hatred, and mental illness, and spread 'post-truth' confusion in the form of many alternative 'realities'.[35] Stories that don't let facts get in the way are much easier to compose than accurate, fact-based stories and — when they're dramatic, and emotionally charged — are much better at grabbing attention. The more it's emotionally charged, the faster it spreads, making it more profitable — and more politically exploitable.

All this has created a new threat to democratic social stability that adds to the older, ongoing threat from gross economic inequality.[2] Among their billions of users the robots, functioning at lightning speed, far outstripping the snail's pace of human moderators, have amplified and expanded the purity spirals, the divisive rhetoric and hatespeech, the misinformed echo chambers, the filter bubbles and preference bubbles, the confusion, and the sheer anger, that's been turbocharging the vicious binary politics and culture wars we see all around us. Ref. 72 describes recent statistical studies of these phenomena.

It's almost as if the social-media giants have been bent on self-destruction. The turbocharged binary politics looks set to replace democracy by brutal autocracy even faster than in the 1920s and 1930s — and this time even at home, even at the nerve centre of the entrepreneurial social media, even in freedom-loving Silicon Valley. Autocracy would destroy the private autonomy of enterprises like Facebook, Meta, and Google. It would destroy the freewheeling, freedom-loving, democratic business environment to which they owe their vast wealth and power. Please don't get me wrong. The social media have their upside and have brought great benefits of many kinds, such as useful networking and videos showing kids how to make and repair things — to say nothing of new ways to *resist* autocracy and to aid high-quality investigative journalism. The threat to democratic social stability was no doubt unintended. But that doesn't make the threat less real, and less imminent.

The social-media technocrats must surely have recognized their peril, whatever their financial bosses might think. The technocrats aren't themselves stupid. There's a good chance, I think, that they're trying to make their robots smarter and less socially destabilizing, difficult though the task may be, and difficult though it may be to overcome the short-term profit incentive for destabilization, and difficult though it may be — on the technical side — to develop robot-assisted moderation at lightning speed. Social scientists are coming up with further ideas that might help,[72] as might new experiments with open-source social media structures like Mastodon, which aim to steer clear of the profit incentive. And the storming of the US Capitol on 6 January 2021 and its threat to turn the USA into an autocracy must, surely, have focused minds, along with the arrival of the new deep-fake disinformation and character-assassination techniques.[21]

Smarter robots could be a game-changer in helping us humans to get smarter, too. They could help us to escape from mindsets instead of being trapped by them. With humans and robots, there's a dystopian mindset that 'they'll take over'. But that's yet another binary trap, another Us Versus Them, based on the 'singularity fallacy' — the unconscious assumption that there's just one kind of intelligence and just one measure of intelligence, putting robots into direct competition with humans. When, instead, humans and robots work together on solving problems, with each playing to their own very *different* strengths, the combination can become much more powerful and exciting than either alone.

A robot can, so to speak, take on the role of a third 'brain hemisphere' to help with problem-solving. Included might even be the problem of maintaining democratic social stability, in all its daunting complexity. Instead of shutting off our thinking, robots could help to open it up. They could, if incentivized in new ways, help us to see things from more than one angle.

An early example was the famous work of the DeepMind team led by Demis Hassabis, with a robot called AlphaGo. In learning to play the combinatorially large game of Go, it discovered winning game patterns that no human had thought of. And now we have its descendant AlphaFold, which in 2020 made a breakthrough toward solving the combinatorially large, and scientifically important, problem of *protein folding*[55] — the problem of deducing the three-dimensional shapes of protein molecules solely from a knowledge of their DNA, hence amino-acid, sequences in cases where the sequence fixes the shape. One of the robots' special strengths is complex pattern recognition within vast sets of possibilities.

The most powerful robots are in some ways, within their limitations, a bit like precocious children. They work by being open to learning. As with human children, we need to get to know them and to get better at teaching them and, above all, better at choosing what to teach them and how best to incentivize them. Should we keep on pushing them to amplify the patterns of social instability, just because it's lucrative in the short term? Is that a smart thing to do? Or should we push them instead to encourage flexible, versatile lateral thinking, and critical thinking, helping to expose things that might surprise us and even make us a bit uncomfortable? Could they help us to become more skilful and adaptable in future? AlphaGo and AlphaFold suggest that the answer is, in principle at least, a resounding *yes*.

We can engage with our own children without knowing the wiring diagrams and patterns of plasticity within their brains. Similarly, we can engage with our robots without knowing what their millions of lines of self-generated computer code look like. And we can get them to help reinforce the more civilized human instincts rather than, as at present, mostly the nastier ones to boost profits. They could for instance do a better job on learning the ever-evolving word patterns and contexts of hatespeech, and catching it, at lightning speed, before it spreads. Such concerns have led to the formation of groups like DeepMind Ethics and Society, the Foundation for Responsible Robotics, and the Center for Humane Technology.

Further discussion can be found under headings such as 'symbiotic future' and 'design fiction'. Design fiction is a technique for using simulation and storytelling techniques to get a variety of people engaged in the design of new technologies — through a hands-on experience of helping to create and try out those technologies, and through storytelling to try to anticipate the possible social effects. A creative toys manufacturer called Twin Science has been getting kids involved in a positive way.

Rutger Bregman's book[11] makes the central point in all this with a famous story, the story of the two wolves. It's dichotomous, but not falsely. "There's a terrible fight going on inside me," says an old man to his grandson. "It's a fight between two wolves. One is evil — greedy, jealous, arrogant, and angry. The other is good — loving, generous, trustworthy, and peaceful. These two wolves are also fighting inside you, and inside everyone else." When the boy asks which wolf will win, the old man replies, "The one you feed."

* * *

As you can see, dear reader, like Demis Hassabis and Rutger Bregman I'm an optimist at heart. However bleak the present may seem, I think there's lots of hope for the future. Everyone needs some kind of faith or hope but, guess what, personal beliefs don't *have* to be puritanical and exclusionary. We don't *have* to take sides. We don't *have* to carry on screaming 'We are right and they are wrong'. We can allow that 'they' are human beings, too. Indeed, group identity needn't itself be rigid. It can be flexible, versatile, and multifarious, as shown for instance by the wonderful experimental studies of Stephen Reicher and co-workers on crowd psychology.[73]

We can work on strengthening our logic-checking and critical thinking — helping us to cope with the rising tide of misinformation, the 'infocalypse' as it's now called.[21] We can cultivate awareness of the traps set by the social-media robots, and awareness of their current aims of getting us addicted, of shutting down thinking, and of highlighting anything that provokes outrage and viral spreading. We can cultivate resistance to the robots' suggested content and filtered newsfeeds, deciding for ourselves what to search for instead. We can cultivate multi-source cross-checking. We can watch out for logical inconsistencies. We can minimize our use of the surveillance machines called digital 'assistants'.[23] We can think twice before entering a metaverse governed by persons unknown. When engaging with our fellow humans, we can cultivate the art of disagreement without hating. We can find space for reasoned dialogue and for the discomfort of uncertainty. We can show respect for others we disagree with.

And, guess what, some of our deepest emotional instincts can help us. Plain old-fashioned compassion and generosity can come into play, welling up from unconscious levels and transcending the game of competitive reciprocal altruism, the game that sees everything as 'deal-making' or zero-sum 'transactional', and sees non-reciprocal altruism as cynical 'virtue signalling'. Deep within us there are genuinely such things as loneliness, empathy, friendship, forgiveness, reconciliation, and even — dare I say it — unconditional love. Love and redemption are forces strongly felt in some of the great epics. All these things have their genetically enabled automata, deep within our unconscious being.

Simplistic evolutionary theory says that such automata cannot exist. They cannot exist, it says, because they're to do with instinctive cooperation rather than competition. But any careful observer can see that they do exist.[2, 10, 11] They must have been selected for, at group level, by the *survival value* of instinctive cooperation and helpfulness. Their ubiquity, their very ordinariness, is attested to in a peculiar way by the fact that they're seldom considered newsworthy.

Of course even love has its dark side. Demagogues love their people and their people love them, until they become tyrants. Insights, including the implausibility of simplistic utopias, such as those on seasteads, or on Mars, go back to the work of Carl Gustav Jung and — as the great novelist Ursula K. Le Guin has reminded us — further back to philosophies such as Taoism that recognize the unconscious intertwining of our very complex dark and light sides, our two wolves, our yin and our yang:

"I would know my shadow and my light, so shall I at last be whole."

In his great oratorio for peace, *A Child of Our Time*, growing out of the 1920s and 1930s, the composer Michael Tippett set those words to some of the most achingly beautiful music ever written.

4

Acausality illusions, and the way perception works

Picture a typical domestic scene. "You interrupted me!" "No, *you* interrupted *me!*"

Such stalemates can arise from the fact that perceived times differ from actual physical times in the outside world, as measured by clocks.[74] I once tested this experimentally by secretly tape-recording a dinner-table conversation. At one point, I was quite sure that my wife had interrupted me, and she was equally sure it had been the other way round. When I listened afterwards to the tape, I discovered to my chagrin that she'd been right. She had started to speak a small fraction of a second before I did.

Musical training includes learning to cope with the differences between perceived times and actual times. For instance, musicians often check themselves with a metronome, a small machine that emits precisely regular clicks. The final performance won't necessarily be metronomic, but practising with a metronome helps to remove inadvertent errors in the fine control of rhythm. "It don't mean a thing if it ain't got that swing…"

There are many other examples. I once heard a radio interviewee recalling how he'd suddenly got into a gunfight: "It all went intuh slowww — motion." Another example is that of the jazz saxophonist, Tony Kofi. At age 16, he fell three stories from a roof-repair job. He describes how he experienced the fall in slow motion, and on the way down had visions of unknown faces and places, and saw himself playing an instrument. It was a life-changing experience that made him into a musician.

A scientist who claims to know that eternal life is impossible has failed to notice that perceived timespans at death might stretch to infinity. That, by the way, is a simple example of the limitations of science. What might or might not happen to perceived time at death is a question outside the scope of science, because it's outside the scope of experiment and observation. It's here that ancient religious teachings show more wisdom, I think, when they say that deathbed compassion and reconciliation are important to us. Perhaps I should add that, as hinted earlier, I'm not myself conventionally religious. I'm an agnostic whose closest approach to the numinous — to the transcendent, to the divine if you will — has been through music.

Some properties of perceived time are very counterintuitive indeed. They've caused much conceptual and philosophical confusion. Besides the 'slow motion' experiences, there are also, for instance, experiences in which the perceived time of an event *precedes* the arrival of the sensory data defining the event, sometimes by as much as several hundred milliseconds. At first sight this seems crazy, and in conflict with the laws of physics. Those laws include the principle that cause precedes effect. But the causality principle in physics refers to actual times in the outside world, not to perceived times. The apparent conflict is a perceptual illusion. I'll call it an 'acausality illusion'.[75]

The existence of acausality illusions — of which music provides outstandingly clear examples, as we'll see shortly — can be understood from the way perception works. And the way perception works is well illustrated by the walking dots animation (Figure 1).

Consider for a moment what the animation tells us. The sensory data are twelve moving dots in a two-dimensional plane. But they're seen by anyone with normal vision as a person walking — as a particular *three-dimensional* motion exhibiting organic change. The invariant elements include the number of dots. Also invariant are the distances, in three-dimensional space, between pairs of locations corresponding to particular pairs of dots. There's no way to make sense of this except to say that the unconscious brain fits to the data an organically changing *internal model*, which represents the three-dimensional motion using an unconscious knowledge of three-dimensional Euclidean geometry.

That by the way is what Kahneman[6] calls a 'fast' cognitive process, something that happens ahead of conscious thought, and outside our volition. Despite knowing that it's only twelve moving dots, we have no choice but to see a person walking.

Such model-fitting has long been recognized by psychologists as an active process involving unconscious prior probabilities, and top-down as well as bottom-up flows of information.[37, 76, 77] The 'top-down' flow comes from the brain's repertoire of internal models, with their unconscious prior probabilities. 'Bottom-up' refers to the incoming data. For the walking dots, the greatest prior probabilities are assigned to a particular class of three-dimensional motions, privileging them over other ways of creating the same two-dimensional dot motion. The active, top-down, model-fitting aspects also show up in neurophysiological studies.[78]

The term pattern-*seeking* is sometimes used to suggest the active nature of the unconscious model-fitting process. So active is our unconscious pattern-seeking that we're prone to what psychologists call pareidolia, seeing patterns in random images. (People see the devil's face in a thundercloud, then form a conspiracy theory that the government covered it up.) For the walking dots, the significant pattern is four-dimensional, involving as it does the time dimension as well as all three space dimensions. Without the animation, one tends to see no more than a bunch of dots.

And what *is* a 'model'? In the sense I'm using the word, it's a partial and approximate representation of reality, or presumed reality. "All models are wrong, but some are useful." And the most useful models are not only representations, but also 'prediction engines' giving a sense of what's likely to happen next, such as whether or not the person walking will take a further step.

Models are made in a variety of ways. They're usually made with symbols of one sort or another. The internal model evoked by the walking dots is made by activating some neural circuitry. Patterns of neural activity are symbols. The objects appearing in video games and virtual-reality simulations are models made of electronic circuits and computer code. Computer code is made of symbols. Children's model boats and houses are made of real materials but are, indeed, models as well as real objects — partial and approximate representations of real boats and houses. Population-genetics models are made of mathematical equations *and* computer code usually. So too are models of photons, of air molecules, of black holes, of lightspeed gravitational ripples, and of jet streams and the ozone hole. Any of these models can be more or less accurate, more or less detailed, and more or less predictive. But they're all partial and approximate. And nearly all of them are made of symbols.

So, ordinary perception, in particular, works by model-fitting. Paradoxical and counterintuitive though it may seem, the thing we perceive *is* — and can only be — the unconsciously fitted internal model. And the model has to be partial and approximate because our neural processing power is finite. The whole thing is counterintuitive because it seems to contradict our subjective visual experience of outside-world reality — as not just self-evidently external, but also as direct, clear-cut, unambiguous, and seemingly exact in many cases.

Indeed, that experience is sometimes called 'veridical' perception, as if it were perfectly accurate. One often has an impression of sharply outlined exactness — with such things as the delicate shape of a bee's wing or a flower petal, the precise geometrical curve of a hanging dewdrop, the sharp edge of the full moon or the sea on a clear day and the magnificence, the sharply defined jaggedness, of snowy mountain peaks against a clear blue sky.[79]

Right now, I'm using the word 'reality' to mean the outside world. Also, I'm assuming that the outside world exists. I'm making that assumption consciously as well as, of course, unconsciously. Notice by the way that 'reality' is itself another dangerously ambiguous word. It's another source of conceptual and philosophical confusion. It's not only that the thing we perceive is called the perceived 'reality', blurring the distinction drawn long ago by Plato and more clearly by Immanuel Kant — the distinction between the thing we perceive and the thing-in-itself, Kant's *Ding an sich*, in the outside world — it's also that there are further ambiguities. Is music real? Is mathematics real? Is our sense of self and gender real? Are love and redemption real? Is religious experience real? There are different kinds of reality belonging to different levels of description. And they differ for different individuals. For me, music is very real and I have an excellent ear for it. When it comes to conventional religion, I'm nearly tone-deaf.

The walking dots remind us that the unconscious model-fitting takes place in time as well as in space. *Perceived times including 'slow motion' experiences are — and can only be — internal model properties.*

And perceived times must make allowance for the brain's finite information-processing rates. That's why, in particular, the existence of acausality illusions is to be expected. In order for the brain to produce a conscious percept from visual or auditory data, many stages and levels of top-down and bottom-up processing are involved. The overall timespans of such processing are familiar from experiments using electroencephalography and magnetoencephalography, which can detect episodes of brain activity. Overall timespans are typically of the order of hundreds of milliseconds. Yet the perceived times of events can have the same 'veridical' character of being clear-cut, unambiguous, and seemingly exact, like the time pips on the radio.

That seeming exactness is a consequence of biological evolution. In toolmaking and in hunting and survival situations, eye–hand–body coordination needs to be as accurate as natural selection can make it. Perceived times need not — and do not — await completion of the brain activity that mediates their perception. Our ancestors survived. We've inherited their timing abilities. We can set a watch to the time pips if we have an FM radio. World-class tennis players time their strokes to within a few milliseconds or thereabouts. World-class musicians work to similar accuracies in the fine control of rhythm and in the most precise ensemble playing. It's more than being metronomic; it's being "on the crest of the rhythm", as some musicians say.

You don't need to be a musician or sportsperson to appreciate the point I'm making. If you and I each tap a plate with a spoon or chopstick, we can easily synchronize a regular rhythm with each other, or synchronize with a metronome, to accuracies far better than hundreds of milliseconds. Accuracies more like tens of milliseconds can be achieved without difficulty. So, it's plain that perceived times — internal model properties — are one thing, while the timings of associated brain-activity events, spread over hundreds of milliseconds, are another thing altogether.

That simple point has been missed again and again in the philosophical and cognitive-sciences literature. In particular, it has caused endless confusion in the debates about consciousness and free will. The interested reader will find further discussion in Ref. 75, but, in brief, the confusion seems to come from an unconscious assumption — which I hope I've shown to be nonsensical — that the perceived 'when' of hitting a ball or taking a decision should be synchronous with the 'when' of some particular brain-activity event.

As soon as that nonsense is blown away, it becomes clear that acausality illusions should exist. And they do exist. The simplest examples come from music — 'the art that is made out of time', as Ursula Le Guin called it in her wonderful novel, *The Dispossessed*. Let's suppose that we refrain from dancing to the music, and that we keep our eyes closed. Then, when we simply listen, the data to which our musical internal models are fitted are the auditory data alone.

I'll focus on Western music. Nearly everyone with normal hearing is familiar, at least unconsciously, with the way Western music works. The familiarity goes back to infancy. Regardless of genre, whether it be commercial jingles, or jazz or folk or rock or classical or whatever — and, by the way, the classical genre includes much film music, for instance John Williams' *Star Wars* — the music depends on precisely timed chord progressions or harmony changes. That's why children learn guitar chords. That's how the *Star Wars* music suddenly goes spooky, at a precise moment after the heroic opening (third audio clip in Figure 11 below).

The musical internal model being fitted to the incoming auditory data keeps track of the times of musical events, including harmony changes. And those times are — and can only be — perceived times, that is, internal model properties.

Figure 5 shows one of the simplest and clearest examples I can find. It's from a classical piano piece that's simple, slow, and serene rather than warlike:

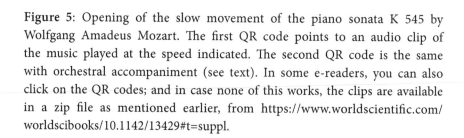

Figure 5: Opening of the slow movement of the piano sonata K 545 by Wolfgang Amadeus Mozart. The first QR code points to an audio clip of the music played at the speed indicated. The second QR code is the same with orchestral accompaniment (see text). In some e-readers, you can also click on the QR codes; and in case none of this works, the clips are available in a zip file as mentioned earlier, from https://www.worldscientific.com/worldscibooks/10.1142/13429#t=suppl.

There are five harmony changes, the third of which is perceived to occur midway through the example, at the time shown by the arrow. Yet, if you stop the playback of the first audio clip at that time, you don't hear any harmony change. Try it! You *can't* then hear any harmony change because the change depends entirely on the next two notes, which are sounded a third and two-thirds of a second after the time of the arrow.

So, in normal playback, the perceived time of the harmony change, at the time of the arrow, precedes by hundreds of milliseconds the arrival of the auditory data defining the change.

That's a clear example of an acausality illusion. It's essential to the way the music works. Almost like the 'veridical' perception of a sharp edge, the harmony change has the subjective force of perceived reality, to the musical ear — the perceived 'reality' of what 'happens' at the time of the arrow.

When I present this example in a lecture, it's sometimes put to me that the perceived harmony change relies on the listener being familiar with the particular piece of music. Having been written by Mozart, the piece is indeed familiar to many classical music lovers. My reply is to present a variant that's *unfamiliar*, with a new harmony change. It starts diverging from Mozart's original just after the time of the arrow, as shown in Figure 6:

Figure 6: As in Figure 5, except that this version departs from Mozart's just after the time of the arrow.

As before, the harmony change depends entirely on the next two notes but, as before, the perceived time of the harmony change — the new and *unfamiliar* harmony change — is at, not after, the time of the arrow. The point is underlined by the way any competent composer or arranger would add an orchestral accompaniment to either example — an accompaniment of the usual kind found in classical piano concertos. Listen to the second audio clip in each figure. The accompaniments change harmony at, not after, the time of the arrow.

I discussed those two examples in greater detail in Ref. 75, with attention to some subtleties in how the two harmony changes work and with reference to the philosophical literature, including Daniel Dennett's multiple-drafts theory of consciousness,[74] which is a way of thinking about perceptual model-fitting in the time dimension.[75]

Just how the brain's neural circuitry carries out the model-fitting is still largely unknown even though the cleverness, complexity and versatility of the process can be appreciated from a huge range of examples.[37, 75–77] Interactions between many brain regions are involved and, in many cases, more than one sensory data stream.

An example of the latter is the McGurk effect in speech perception. *Visual* data from lip-reading can cause changes in the perceived *sounds* of phonemes. For instance, the sound 'baa' is often perceived as 'daa' when watching the lips of someone saying 'gaa'. The phoneme model is being fitted multi-modally — simultaneously to more than one sensory data stream, in this case visual and auditory. The brain often takes 'daa' as the best fit to the slightly conflicting data, the auditory 'baa' and the visual 'gaa'.

The Ramachandran–Hirstein 'phantom nose illusion' — which can be demonstrated without special equipment[37] — produces a striking distortion of one's perceived body image, a nose elongation well beyond Pinocchio's or Cyrano de Bergerac's. It's produced by a simple manipulation of tactile and proprioceptive data. They're the data feeding into the internal model that mediates the body image, including the proprioceptive data from receptors such as muscle spindles sensing limb positions.

And what's this so-called body image? Well, the brain's unconscious internal models must include a *self-model* — a partial and approximate representation of one's self and one's body in its surroundings. Plainly one needs a self-model, if only to be well oriented in one's surroundings and to distinguish oneself from others. "Hey — *you're* treading on *my* toe."

There's been philosophical confusion on this point too. Such a self-model must be possessed by any animal. Without it, neither a leopard nor its prey would have a chance of surviving. Nor would a bird, nor a bee, nor a fish. Any animal needs to be well oriented in its surroundings and to be able to distinguish itself from others. Yet the cognitive and philosophy-of-mind literature sometimes conflates 'having a self-model', on the one hand, with 'being conscious' on the other — two separate things.

The self-model and body image are sometimes activated in an unusual way to produce 'out-of-body' experiences. The brain's model of the body in its surroundings is of course a three-dimensional model, potentially able to generate a view from any angle. The point is illustrated by the story of the artist Franco Magnani, described in one of Oliver Sacks' books.[8] From memory, Magnani made accurate paintings of his childhood village of Pontito viewed from many angles, including aerial views.

Another source of confusion is the idea that symbolic representation came into existence only recently, as part of the human cognitive revolution near the start of the Upper Palaeolithic.[12, 62] The point is missed that leopards and their prey can perceive things and therefore have internal models. So do birds, bees, and fish. Their internal models, like ours, are — and can only be — unconscious symbolic representations. Again, patterns of neural activity are symbols. Symbolic representation is far more ancient — by hundreds of millions of years — than is sometimes supposed.

The use of echolocation by bats, whales and dolphins is a variation on the same theme. For bats, too, the perceived reality must be the internal model — not the echoes themselves, but a symbolic representation of the bat in its surroundings. It must work in much the same way as our vision, except that the bat provides its own illumination. To start answering the famous question 'what is it like to be a bat', we could do worse than imagine seeing in the dark with a flashing floodlight, whose rate of flashing can be increased at will.

And what is it like to be an octopus? Like any other animal, an octopus needs to be well oriented in its surroundings. So, regardless of brain anatomy, it needs a single self-model and an accompanying perception of 'self' — not the absurdity of eight or nine separate 'selves' as has sometimes been suggested. The anatomy of an octopus, whose brain is spread out across its eight arms, is irrelevant to the question. An entirely different question is where to focus your attention. For us humans, we can focus on what our left fingers and right toes are doing. So I daresay an octopus can focus on what its fifth and seventh arms are doing.

And what of the brain's two hemispheres? Here I must defer to McGilchrist[36] and Ramachandran,[37] who in their different ways offer a rich depth of understanding coming from neuroscience and neuropsychiatry, far transcending the superficialities of popular culture. For present purposes, McGilchrist's key point is that having two hemispheres is evolutionarily ancient. Even fish have them. The two hemispheres may have begun with bilateral symmetry in primitive vertebrates, but then evolved in different directions. If so, it would be a good example of how neutral genomic changes can later become adaptive.[18]

A good reason to expect such bilateral differentiation, McGilchrist argues, is that survival is helped by having two different styles of perception. They might be called holistic and fragmented. The evidence shows that the first, holistic style is a specialty of the right hemisphere, and the second is a specialty of the left, or *vice versa* in a minority of people.

If you're a pigeon who spots some small objects lying on the ground, then you want to know whether they are, for instance, grains of sand or nutritious seeds. That's the left hemisphere's job. It has a style of model-fitting, and a repertoire of internal models, that's suited to a fragmented, dissected view of the environment, with an intense focus on fine details. The left hemisphere can't see the wood for the trees. Or, more accurately, it can't even see a single tree but only, at best, leaves, twigs or buds (which, by the way, might be good to eat). One can begin to see why the left hemisphere is more prone to unconscious mindsets.

But suppose that you, the pigeon, are busy sorting out seeds from sand grains and that there's a peculiar flicker in your peripheral vision. Suddenly, there's a feeling that something is amiss. You glance upward just in time to see a bird of prey descending and you abandon your seeds in a flash! That kind of perception is the right hemisphere's job. The right hemisphere has a very different repertoire of internal models, holistic rather than dissected. They're fuzzier and vaguer, but with a surer sense of overall spatial relations, such as your body in its surroundings. They're capable of superfast deployment. The fuzziness, ignoring fine detail, makes for speed when coping with the unexpected. Ref. 36 gives many more examples.

Ref. 37 tells us that another of the right hemisphere's jobs is to watch out for inconsistencies between incoming data and internal models, including any model that's currently active in the left hemisphere. When the data contradict the model, the left hemisphere has a tendency to reject the data and cling to the model — to be trapped in a delusional mindset. "Don't distract me; I'm trying to concentrate!" Brain scans show a small part of the right hemisphere that detects such inconsistencies or discrepancies. The right hemisphere can interrupt the left with a wordless "Look out, you're making a mistake!" If the right hemisphere's discrepancy detector is damaged, severe delusional mindsets such as anosognosia can result.

Ref. 36 points out that the right hemisphere is involved in many subtle and sophisticated games, such as playing with the metaphors that permeate language or, one might even say, that *mediate* language. So, the popular-cultural idea that language is all in the left hemisphere misses many of the deeper aspects of language.

And what of combinatorial largeness? Perhaps the point is obvious. For instance there's a combinatorially large number of possible visual scenes, and of possible assemblages of internal models to fit them. Even so simple a thing as a chain with 10 different links can be assembled in 3,628,800 different ways, and with 100 different links in approximately 10 to the power 158 different ways, 1 followed by 158 zeros. Neither we nor any other organism can afford to deal with all the possibilities, even unconsciously. Visual-system processes such as early-stage edge detection[77, 80] and the unconscious perceptual grouping studied by the Gestalt psychologists[76] as with the two groups in dot patterns like •• ••• provide glimpses of how the vast combinatorial tree of possibilities is pruned by our extraordinary model-fitting apparatus — the number of possibilities cut down at lightning speed and ahead of conscious thought.

Perceptual grouping works in time as well as in space, as for instance with the four-note groups in Figures 5 and 6. Such grouping in perceived time was adumbrated long ago in the thinking of the philosopher Henri-Louis Bergson, published in 1889, predating the work of the Gestalt psychologists in the early twentieth century. It's part of what's involved in the acausality illusions illustrated in Figures 5 and 6.

Perceptual grouping occurs in word patterns too, of course, as in "Eggs must be marked with the date on which they were laid by the farmer."

And what of science itself? What about all those mathematical and computer-coded models of population genetics and of photons, of air molecules, of black holes, of lightspeed gravitational ripples, of jet streams and the ozone hole, of invisible pandemic spreading and of the myriad other entities we deal with in science? Could it be that science itself is always about finding useful models that fit data from the outside world, and never about finding Veridical Absolute Truth? Can science be a quest for truth even if the truth is never absolute? The following chapter will argue that the answer to the last two questions is an emphatic *yes*.

Take, for instance, the physicists' Holy Grail, the ultimate model or 'Theory of Everything' that combines gravity and quantum effects. Suppose we were to find a candidate ultimate model, consistent with everything we know today and described by a single, self-consistent set of mathematical equations. It would still be impossible to test it at infinite accuracy, in an infinite number of cases, and in all parts of the Universe or Universes, past, present, and future. Within a smaller, but still wide, domain of applicability, one might achieve superlative scientific confidence, with many accurate cross-checks including new predictions subsequently verified. The model might be described by equations of consummate beauty. And that would be wonderful. But in principle there'd be no way to be Absolutely Certain that it's Absolutely Correct, Absolutely Accurate, and Applicable to Absolutely Everything.

Come to think of it, isn't that kind of obvious?

5

What is science?

Where then does all this lead? I think it leads to something simple yet profound and far-reaching. Let me put it a touch provocatively. I'd like to replace all those books on the philosophy of science by just one simple statement. It not only says what science is, in the most fundamental possible way, but it also clarifies the power and limitations of science. It says that *science is an extension of ordinary perception*, meaning perception of outside-world reality. Like ordinary perception, science fits models to data from the outside world.

If that sounds glib and superficial to you, dear reader, then all I ask is that you think again about the sheer wonder of so-called ordinary perception. It too has its power and its limitations, and its fathomless subtleties, agonized over by generations of philosophers.

Both science and ordinary perception work by fitting models — symbolic representations — to data coming in from the outside world. Both science and ordinary perception must assume that the outside world exists, because it can't be proven absolutely. Models, and assemblages and hierarchies of models, schemas or schemata as they're sometimes called, are partial and approximate representations, or candidate representations, of outside-world reality. Those representations can be anything from superlatively accurate and strongly predictive to completely erroneous — like the phlogiston theory of combustion, the microwave theory of COVID-19, and ordinary hallucinations.

The walking dots animation of Figure 1 points to the tip of a vast iceberg, a hierarchy of unconscious internal models and model components starting with the three-dimensional motion itself, but extending all the way to the precise manner of walking and the associated psychological and emotional subtleties. The main difference between science and so-called ordinary perception lies in the range of models used, in the data to be fitted, and in a more explicit focus, by science, on estimating degrees of uncertainty.

In science today we can harness the power of Bayesian causality theory[22] to fit sophisticated models to vast datasets in a logically self-consistent way, using the probabilistic 'do' operator to represent the actions of an experimenter. The theory also has a natural way of dealing with uncertainty. Problems of the most daunting complexity are thus beginning to be tractable. Examples include the complex biomolecular circuitry that switches genes on and off,[3, 9, 18] and the interplay between small-scale ocean eddies and global-scale circulations and weather systems.[81]

Notice that all our ways — scientific and ordinary — of perceiving the outside world are 'theory-laden' as is sometimes said. One might also say 'prior-probability-laden'. It's a necessary aspect of any model-fitting process. Consciously or unconsciously, one has to begin somewhere when selecting models to fit. Consciously or unconsciously, one has to propose some pattern of cause and effect before it can be tested against data.[22] Some postmodernist[35] philosophers such as Paul Feyerabend have claimed that scientific knowledge is mere opinion, just because it's theory-laden. The point is missed that some models fit better than others. And some have more predictive power than others. And some are *a priori* more plausible than others, with more cross-checks to boost their prior probabilities. And some are simpler and more widely applicable than others.

Take for instance Newton's and Einstein's models of gravity. Both are partial and approximate representations of reality even though superlatively accurate, superlatively simple, superlatively predictive, and repeatedly cross-checked in countless ways within their very wide domains of applicability — Einstein's still wider than Newton's because it includes, for instance, the orbital decay and merging of pairs of black holes or neutron stars and the resulting lightspeed gravitational ripples already mentioned, also called gravitational waves. They are ripples in the structure of spacetime itself, and were first observed in 2015 by the famous 'LIGO' detectors.[16] Observation of the gravitational ripples has provided yet another cross-check on Einstein's model — which predicted them over a century ago — and has opened a new window on the Universe.

Both models are not only simple but also mathematically beautiful. And their high accuracies and predictive powers are crucial, for instance, to all our achievements in space science and space travel. The way a spacecraft moves isn't a matter of mere opinion.

All this has to do, then, with cross-checking — including the checking of model predictions in new situations — and with data quality, accuracy of fit, beauty and economy of modelling, and domain of applicability. It's never about Absolute Truth and Absolute Proof. Nor is it about uniqueness of model choice. There might be two or more alternative models that work equally well. They might have comparable simplicity and accuracy and offer complementary, and equally powerful, insights into the workings of outside-world reality. The possibility of such non-uniqueness has recently been emphasized in parts of the literature on biological evolution where it's sometimes, rather loosely, called 'equivalence' between models[43] and is being used as part of efforts to view things from more than one angle, and to escape from the mindsets of simplistic evolutionary theory.

The possibility of non-uniqueness is troublesome for believers in Absolute Truth. And it's much agonized over in the philosophical literature, under headings such as 'incommensurability'. However, as I keep saying, not even the existence of the outside world can be *proven* absolutely. It has to be assumed. Both science and ordinary perception proceed on that assumption. The justification is no more and no less than our experience that model-fitting works, again and again — never perfectly, but often well enough to gain our respect, especially when it makes useful predictions.

If you observe a rhinoceros charging towards you, then it's probably a good idea to jump out of the way even though your observations are, unconsciously, theory-laden — theory-laden from hundreds of millions of years of evolution — and even though there's no absolute proof that the rhinoceros exists. Even a postmodernist might jump out of the way. And Einstein's gravitational ripples gain our respect not only for the technical triumph of observing them but also because the merging black holes emit a very specific wave pattern, closely matching the details of what's computed from Einstein's equations when the black holes have particular masses and spins.[16] So it's reasonable to suppose that, as with rhinoceroses, black holes exist and behave in certain ways. That's reasonable even though our knowledge of them is theory-laden.

Another tendency among believers in Absolute Truth seems to be a tendency to blur Kant's distinction, the distinction between outside-world reality and our models of it. A model is viewed as the *same thing as* the outside-world reality. The late Edwin T. Jaynes aptly called this conflation the 'mind-projection fallacy'.[82, 83] It's another variation on the theme of veridical perception.[79]

Mind projection sometimes takes extreme forms in physics and cosmology. It can generate a transcendent vision of Absolute Truth in which the entire Universe is seen as a single, unique mathematical object of supreme beauty, giving us a 'Theory of Everything' — *the* unique, exact Answer to Everything — residing within that timeless ultimate Reality, the Platonic world of perfect mathematical forms.[82, 84] Alleluia! (In Chapter 6, though, I'll dare to propose a very different idea of what the Platonic world is, and where it comes from.)

Because the model-fitting works better in some cases than in others, there are always considerations of how well it is working, involving a balance of probabilities. We must always consider how many independent cross-checks have been done and to what accuracies. For Einstein's equations, the detection of gravitational ripples from merging black holes didn't suddenly 'prove' Einstein's theory — as some journalists had it — but instead just added a new independent cross-check, though an awesomely beautiful one. There were many earlier cross-checks including a delicate, but accurate, check from what's called the Hulse–Taylor binary. Two neutron stars circle each other, generating gravitational ripples. The consequent orbital decay was calculated from Einstein's equations and confirmed observationally, using pulsar radio emissions from one of the stars.

If you can both hear and see the charging rhinoceros and if your feet feel the ground shaking in synchrony, then you have some independent cross-checks. You're checking a single internal model, unconsciously of course, against three independent sensory data streams. With so much cross-checking, it's a good idea to accept the perceived reality as a practical certainty, theory-laden though it is.

We do it all the time. Think what's involved in riding a bicycle, or in playing tennis, or in pouring a glass of wine. But the *perceived reality* is still the *internal model* within your unconscious brain, paradoxical though that may seem. And, again, the outside world is something whose existence must be assumed.

One reason I keep banging on about these issues is the quagmire of philosophical confusion that has long surrounded them. It includes but also predates the postmodernist commentaries and 'science wars', and never seems to go away.[75, 79, 85] Back in the 1920s, the famous Vienna Circle of leading scientists and philosophers assumed that there are such things as direct or veridical observations — 'empirical' information sharply, absolutely, and infallibly distinct from models. Others have argued that the entire outside world is an illusion, subjective experience being the only reality. Ref. 85 suggests that perception might be a peculiar mixture of veridical directness on the one hand, and some kind of theory-laden model fitting on the other. But none of this helps! Like the obsession with uniqueness and absolute proof — and absolute truth versus absolute falsehood — it just gets us into a muddle.

Journalists often press us to say whether something is scientifically 'proven' or not even though, as Karl Popper pointed out long ago, that's yet another false dichotomy — another perilous binary button. I think professional codes of conduct for scientists should reject such absolutism. We should talk instead about the balance of probabilities and the degree of scientific confidence.

Many scientists do that already. *Journalist to scientist*: "Professor So-and-so, is that vaccine safe, or not? You must answer yes or no." *Scientist to journalist*: "No I mustn't! There's no such thing as absolute safety. What I *can* say is that the vaccine is safe for everyone except about one in a million, on current evidence. And your chance of serious illness without the vaccine is thousands of times greater."

Let me come clean. I admit to having had my own visitations of the absolute — my own epiphanies, eurekas, and alleluias. But as a professional scientist I wouldn't exhibit them in public, at least not as Absolute Truths. They should be for consenting adults in private, an emotional resource to power our research efforts, not something for scientists to air in public. I think most of my colleagues would agree. We don't want to be lumped in with all those cranks, zealots, and conspiracy theorists who believe "in a single truth and in being the possessor thereof." And again, even if a candidate 'Theory of Everything', so called, were to be discovered one day, the most that science could ever say is that it fits a large but finite dataset to within a small but finite experimental error.

Consider again the walking dots animation. Instinctively, we *feel sure* that we're looking at a person walking. "Hey, that's a person walking. What could be more obvious?" Yet the animation might not come from a person walking at all. The prior probabilities, the unconscious selection of models to fit, the theory-ladenness, might be wrong. The twelve moving dots might have been produced in some other way. The dots might 'really' be moving in a two-dimensional plane, or three-dimensionally in any number of ways. And even our charging rhinoceros might, just might, be a hallucination. As scientists we *always* need to consider the balance of probabilities, trying to get as many cross-checks as possible and trying to reach well-informed judgements about the degree of scientific confidence. That's what was done with the ozone-hole work in which I was involved professionally, which defeated the ozone disinformers. That's what was done in the 1920s with the discovery and testing of quantum theory, where *nothing* is obvious!

There is, of course, a serious difficulty here, on the level of communication skills, communication tactics, and lucidity principles. We do need quick ways to express extremely high confidence, such as confidence in the sun rising tomorrow, and confidence that the sunrise won't be caused by a cock crowing. We don't want to waste time on such things when confronted with far greater uncertainties. Scientific research is like driving in the fog, straining to see the twists and turns ahead. So there's a temptation to use terms like 'proof' and 'proven' just as a shorthand to indicate things to which we attribute practical certainty, things we shouldn't be worrying about when trying to see other things through the fog.

But because of all the philosophical confusion I still think it preferable in public to avoid terms like 'proof' or 'proven', or even 'settled', and instead try to use a more nuanced range of terms like 'practically certain', 'well established', 'well verified', 'highly probable', 'very well checked', and so on, when we feel that strong statements are justifiable in the current state of knowledge. Such terms sound less final and less absolutist, especially when we're explicit about the strength of the evidence and the variety of cross-checks. I'll try to set a good example in the *Postlude* on climate.

Also to be avoided, I think, is that other binary button 'fact versus theory'. Pressing it plays straight into the hands of the professional disinformers, those well-resourced masters of dichotomized information warfare who work to discredit good science whenever it's seen as threatening short-term profits. The 'fact versus theory' button gives them a ready-made framing tactic,[41] unconsciously paralleling 'good versus bad' and 'like versus dislike'.

* * *

I want to return briefly to the fact — the practical certainty, I mean — that what's hopelessly complex at one level can be simple, or at least understandable, at another.

Multiple levels of description are not only basic to science but also, unconsciously, basic to ordinary perception. They're natural. They're basic to how our brains work. Straight away, our brains' left and right hemispheres give us at least two levels of description, a lower level that dissects fine details, and a more holistic higher level.[36] And neuroscience has revealed a variety of specialized internal models or model components that symbolically represent different aspects of outside-world reality. In the case of vision, there are separate model components representing not only fine detail on the one hand, and overall spatial relations on the other but also, for instance, motion and colour. Damage to a part of the brain dealing with motion can produce visual experiences like successions of snapshots or frozen scenes — merely a nuisance if you're trying to pour your tea, but very dangerous if you're trying to cross the road.[77] Other kinds of brain damage can produce, for instance, colours floating around by themselves and unattached to objects.[85]

The biological sciences well illustrate the need to consider multiple levels of description, and multiple modes of description. I've already mentioned molecular-biological regulatory networks, or biomolecular circuits.[3, 9, 18] They depend on highly specific interactions between a variety of molecules including DNA and protein molecules. Shape-changing protein molecules called 'allosteric enzymes' function within biomolecular circuits like transistors within electronic circuits. Biomolecular circuits and their actions, such as switching genes on and off, are impossible to recognize from lower levels such as the level of genes alone, still less from the levels of chemical bonds and bond strengths within thermally agitated molecules, jiggling around and bumping into each other on timescales of trillionths of a second, and from the still lower levels of atoms, electrons, atomic nuclei, and quarks.

And again, there are many higher levels of description within the hierarchy — level upon level, with causal arrows pointing both downward and upward. There are biomolecular circuits and assemblies of such circuits, going up to the levels of archaea, bacteria and their communities, of yeasts, of multicellular organisms, of niche construction and whole ecosystems, and of ourselves and our

families, communities, nations, cyberspace, and the entire planet —
which Newton treated as a point mass.

None of this would need saying were it not for the persistence —
even today — of an extreme-reductionist view of science saying, or
assuming, that looking for the lowest possible level and for atomistic
'units' such as quarks, or atoms, or genes, gives us the Answer to
Everything and is therefore *the only useful angle* from which to view
a problem. Some of the disputes about biological evolution seem to
have been disputes about 'the' unit of selection[58] — as if such a thing
could, or should, be uniquely identified within the actual complexity
of multi-level selection. Yes, in many cases reductionism can be
enormously useful; but no, it isn't the Answer to Everything!

In some scientific problems, including those I've worked on
myself, the most useful models aren't at all atomistic. In fluid
dynamics we use accurate 'continuum-mechanics' models in
which highly nonlocal, indeed long-range, interactions are crucial.
They're mediated by the pressure field. They're a crucial part of, for
instance, how birds, bees, and aircraft stay aloft, how a jet stream
can circumscribe and contain the ozone hole, and how waves and
vortices interact.[31, 86]

McGilchrist's work tells us that extreme reductionism comes
from our brains' left hemispheres. It is indeed a highly dissected
view of things. His book[36] can be read as a passionate appeal for
more pluralism — for more of Max Born's loosening of thinking —
for the more powerful, in-depth understanding that can come from
looking at things on more than one level and from more than one
viewpoint, while respecting the evidence. Such understanding
requires a better collaboration, says McGilchrist, between our
garrulous and domineering left hemispheres and our quieter, indeed
wordless, but also passionate, right hemispheres (or *vice versa* in a
minority of people).

Surely, then, professional codes of conduct for scientists — to say nothing of lucidity principles as such — should encourage us to be *more explicit than we feel necessary* regarding, in particular, which level or levels of description we're talking about. And when the levels of description aren't clear, or when the questions asked are 'wicked questions' having no clear meaning at all, still less any clear answer, it would help to say so.

Such an approach might also be helpful when confronted by all the confusion, and wicked questions, about consciousness and free will. I want to stay off those topics — having already had a go at them in Ref. 75 — except to say that some of the confusion seems to come first from not recognizing the existence of acausality illusions, and second from conflating different levels of description. I like the aphorism that "free will is a biological illusion but a social reality". There's no conflict between the two statements once they're recognized as belonging to different levels of description.[87] The same goes for statements like "This individual is biologically a man, but mentally and socially a woman."

And once again we're reminded of the ambiguity, the context-dependence, and the multi-level aspects of the word 'reality'. I ask again, is music real? Is religious experience real? Is mathematics real? Is our DNA real? Is the outside world real? Is our sense of self and gender real? For me, at least, they're all real but in different senses, relevant at different levels, that keep on being confused with one another. And one of life's realities is that pragmatic functioning at the mental and social levels depends on accepting our sense of self and gender — our internal self-model — as an entity having, or seeing itself as having, free will or volition or agency as it's variously called. It wouldn't do to be able to commit murder and then, like a present-day Hamlet, to say to the jury "T'wasn't me, t'was my selfish genes did it."

6

Music, mathematics, and the Platonic

The walking dots show that we have unconscious Euclidean geometry. We also have unconscious calculus. Calculus illustrates one of the most wonderful and surprising things about mathematics, namely that a pattern helpful in solving one problem — such as predicting the path of a tennis ball — later becomes key to solving different problems at first sight unrelated, such as understanding electric currents and power grids, or jet streams.

Calculus is the mathematics of continuous change. For instance, it deals with objects like those in Figure 7:

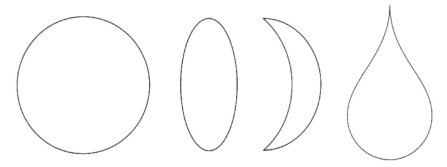

Figure 7: Some smooth 'mathematical' curves.

Such objects are made of smooth curves — pathways whose direction changes continuously, the curves that everyone calls 'mathematical'. They include perfect circles, ellipses, and portions thereof, among countless other examples. A straight line is the special case having zero rate of change of direction. A circle has a constant rate of change of direction, and an ellipse has a rate of change that's itself changing, and so on.

Such 'Platonic objects', as I'll call them, are of special interest to the unconscious brain. Whenever one sees natural phenomena exhibiting what look like straight lines or smooth curves, such as the edge of the sea on a clear day, or the edge of the full moon, or the shape of a hanging dewdrop, they tend to excite our sense of something special, and beautiful. So do the great pillars of the Parthenon, and the smooth curves of the Sydney Opera House seen from a distance. We feel their shapes as resonating with something 'already there'.

Plato felt that the world of such 'mathematical' objects, shapes, or forms, and the many other beautiful entities found in mathematics, is in some sense a world more real than the outside world with its commonplace messiness. He felt his world of perfect mathematical forms to be something timeless — something already there and always there.

My heart is with Plato here, on an emotional level. When the shapes, or forms, look truly perfect, they can excite a sense of great wonder and mystery. So too can the mathematical equations describing them. How can such immutable perfection exist at all, and why do we find it awesome?

Indeed, so powerful is our unconscious interest in such perfection that we see smooth curves *even when they're not actually present in the incoming visual data*. For instance, we see them in the form of what psychologists call 'illusory contours'. Figure 8 is an example. If you stare at the inner edges of the messy black marks for several seconds, and if you have normal vision, you'll see an exquisitely smooth curve joining them:

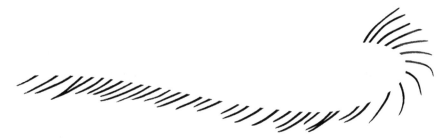

Figure 8: An illusory contour. To see it, stare at the inner edges of the black marks.

That curve is not present on the screen or on the paper. It's constructed by your visual system. To construct it, the system unconsciously solves a problem in calculus — in the branch of it called the calculus of variations. The problem is to consider all the possible curves that can be fitted to the inner edges of the black marks, and to pick out the curve that's as smooth as possible, in a sense to be specified. The smoothness is specified using some combination of rates of change of direction, and rates of change of rates of change, and so on, averaged along each curve. The smaller the rates of change, the greater the smoothness. So we have an unconscious calculus of variations. And that in turn gets us closer to some of the deepest parts of theoretical physics, as we'll see shortly.

The existence of the Platonic world glimpsed in Figures 7 and 8 is no surprise from an evolutionary perspective. It is, indeed, already there. It's already there in the sense of being evolutionarily ancient — something that comes to us through genetic memory and the automata that it enables — genetically enabled automata that can self-organize or self-assemble into, among other things, the special kinds of symbolic representation that correspond to Platonic objects.

Over vast stretches of time, natural selection has put the unconscious brain under pressure to make its model-fitting processes as simple as the data allow. That requires a repertoire of internal model components that are as simple as possible. Some of those components are Platonic objects, smooth curves, or portions of smooth curves or, rather, their internal symbolic representations. Please remember that actual or latent patterns of neural activity are symbols — and we are now talking about mathematical symbols — even though we don't yet have the ability to read them directly from the brain's neural networks.

A perfect circle, then, is a Platonic object simply because it's simple. And the illusory contour in Figure 8 shows that the brain's model-fitting process assigns the highest prior probabilities to models representing objects with the simplest possible outlines consistent with the data, in this case a light-coloured object with a smooth outline sitting in front of some smaller dark objects. That's part of how the visual system distinguishes an object from its background, an important part of making sense of the visual scene.[8, 76, 77, 80] Making sense of the visual scene has been crucial to survival for hundreds of millions of years — crucial to navigation, crucial to finding mates, and crucial to eating and not being eaten. Many of the objects to be distinguished have outlines that are more or less smooth. They range from distant hills down to fruit and leaves, tusks and antlers, and teeth and claws.

"We see smooth curves even when they're not actually present." Look closely at Figure 7. *None* of the Platonic objects we see are actually present in the figure. Take the circle on the left, or ellipse as it may appear on some screens. It's actually more complex. With a magnifying glass, one can see small-scale roughness. Zooming in more and more, one begins to see more and more detail. One can imagine zooming in to the atomic, nuclear, and subnuclear scales. Long before that, one encounters the finite scales of the retinal cells in our eyes. Model-fitting is partial and approximate. What's complex at one level can be simple at another. Perfectly smooth curves are things belonging not to the incoming sensory data but rather — I emphasize again — to the unconscious brain's repertoire of internal model components.

So I'm suggesting that the Platonic world is very different from what Plato, and others, seem to have imagined.[82, 84] Rather than being timeless, it's only hundreds of millions of years old. Rather than being external to us, it's very *internal*. It's something arising from natural selection. It's part of the unconscious mathematics we need in order to survive. But to me that's still wonderful, indeed even more wonderful because, for one thing, it makes a lot more sense.

* * *

The calculus of variations is a gateway to some of the deepest parts of theoretical physics. That's because it leads to Noether's theorem. The theorem depends on writing the equations of physics in what's called 'variational' form. That's a form allowing the calculus of variations to be used. It's Richard Feynman's own example of things that are mathematically equivalent but, as he said, "psychologically very different".

Think of playing tennis on the Moon. Air resistance is assumed negligible. After being hit, the ball moves solely under gravity. One way to model such motion is to use calculus in the form of Newton's equations. They deal with the moment-to-moment rates of change of quantities like the position and velocity of the tennis ball. In our lunar example, solving those equations produces a pathway for the tennis ball in the form of a smooth curve, very close to what's called a parabola. Since the equations describe what happens from moment to moment, one must construct the curve bit by bit, starting at one end.

However — and this might seem surprising — the same smooth curve can also be constructed as the solution to a variational problem. It's a problem more like that of Figure 8 because it treats the curve holistically, rather than bit by bit. It deals with all parts of the curve simultaneously. That's psychologically very different indeed.

One considers all possible curves beginning and ending at a given pair of points. Instead of finding the smoothest of those curves, however, the problem is to find the curve having the smallest value of another property, quite different from any measure of roughness or smoothness. That property is the time average, along the whole curve, of the velocity squared minus twice the gravitational altitude, or gravitational energy per unit mass. In order for the problem to make sense one has to specify a fixed travel time as well as fixed end points in space. Otherwise, the velocity squared could be anything at all. Apart from a multiplicative constant, the time average in question is known in physics as the 'action integral' for the problem, or 'the action' for brevity.

If one solves this variational problem, minimizing the action over all possible curves, then one ends up with exactly the same curve as one gets from solving Newton's equations. Even though psychologically very different, the variational problem is *mathematically equivalent* to Newton's equations.

Using the standard methods of the calculus of variations — the same methods as for Figure 8 — one can manipulate the mathematical symbols to show that the equivalence holds in all possible cases. Mathematics does indeed handle many possibilities at once. For the reader wanting more detail I'd recommend the marvellous discussion in Feynman's *Lectures in Physics*, which goes on to discuss the deep significance for quantum theory.[88]

Once one has the problem in variational form, one can apply Noether's theorem. The theorem tells us that, in each case of tennis-ball motion, we have organic change in the abstract sense I've defined it. There are conserved quantities, including an invariant called the total energy. They stay the same while the tennis ball changes its position and velocity. The total energy is proportional to the velocity squared *plus* twice the gravitational altitude.

Noether's theorem tells us that, in particular, the invariance of the tennis ball's total energy follows from a property of the action called time-symmetry. That property says that the action has no dependence on the starting time, but depends only on the travel time, the specified time interval between starting and finishing. Starting at a different time makes no difference to the problem. For our tennis ball this time-symmetry condition is satisfied insofar as the gravitational field can be considered steady, i.e. not changing with time.

The invariant quantities revealed by Noether's theorem become more and more important as we deal with problems that are more and more complicated. Invariant total energies give us a key to showing why, for instance, it's a waste of time trying to build a complicated perpetual-motion machine, or a machine to supply your household electricity with no energy input.

The invariants are crucial to large parts of theoretical physics. Fluid dynamics, electromagnetism, quantum theory, and particle physics are included. The invariants in all these cases can be found by the systematic use of Noether's theorem. All that needs changing from case to case is the formula for the action, together with the space in which it's calculated. The mathematical framework stays the same. There are a number of different types of invariants because there are a number of different types of symmetry, depending on the problem under consideration. The symmetries are described by another powerful branch of mathematics called group theory.

Invariants in fluid dynamics have an important role, for instance, in understanding jet streams and the ozone hole, and the chaotic dynamics of eddies in the atmosphere and oceans.[31]

* * *

Music has its own Platonic objects. Prominent among them are the special sets of musical pitches called harmonic series. An example is shown in Figure 9, where the leftmost QR code points to an audio clip sounding the pitches one after another:

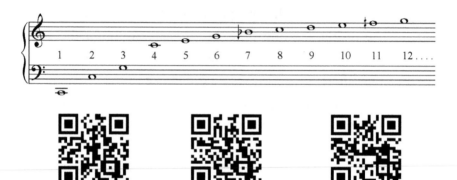

Figure 9: A musical harmonic series. The first QR code, on the left, points to an audio clip sounding these harmonic-series pitches one after another. The second and third QR codes point to audio clips of some of the pitches sounded together (see text). In some e-readers you can also click on the QR codes; and in case none of this works, the clips are available in the zip file mentioned earlier, from https://www.worldscientific.com/worldscibooks/10.1142/13429#t=suppl.

The first and lowest pitch, called the 'fundamental' or 'first harmonic', corresponds in this case to a vibration frequency 65.4Hz (65.4 hertz, i.e. cycles per second). The second pitch or harmonic corresponds to twice this, 130.8Hz, and the third to three times, 196.2Hz, and so on. The fundamental pitch and its octave harmonics, the 2nd, 4th, 8th, and so on all have the same musical name C or Doh. If you happen to have a tunable electronic keyboard and would like to tune it to agree with the harmonic series, then you need to sharpen the 3rd, 6th, and 12th harmonics by 2 cents (2/100 of a semitone) and the 9th by 4 cents — these differences are barely audible — but also to flatten the 5th and 10th by

Caption continues on facing page

14 cents (audible to a good musical ear), the 7th by 31 cents, and the 11th by 49 cents, relative to B flat and F sharp. The last two changes are plainly audible to just about anyone. The differences arise from the fact that the standard keyboard tuning, called 'equal temperament', divides the octave into twelve exactly equal semitones each with frequency ratio 1.059463.., the twelfth root of 2. Equal temperament is musically useful because of a peculiar accident of arithmetic. That's the tiny, practically inaudible 2-cent difference between the 3rd harmonic and its equal-tempered approximation, whose frequency is 1.49831, the 7th power of the semitone frequency ratio — very nearly 3/2 — times the frequency of the 2nd harmonic. The tuning of the 3rd harmonic pitch, relative to the 2nd, has critical importance in most genres of music.

The defining property of a harmonic series is that the pitches correspond to vibration frequencies equal to the lowest frequency, in this case 65.4 hertz (65.4 cycles per second), multiplied by a whole number such as 1, 2, 3, etc.

A harmonic series is a 'Platonic object' in just the same sense as before, something that's evolutionarily ancient and of special interest to the unconscious brain. How can that be? The answer will emerge shortly, when we consider how hearing works. And it will expose more connections between music and mathematics. But first, please be sure to listen to the sounds themselves (first or leftmost audio clip in Figure 9). Do they not hint at something special and beautiful? Something that could divert you, and Plato, from mundane messiness? Hints of fairy horn calls, perhaps? However they strike you, these sounds are special to the musical brain.

Also special are subsets of these sounds played together. For instance, if those numbered 4, 5, and 6 are played together — they are called the 4th, 5th, and 6th harmonic pitches — then we hear the familiar sound of what musicians call a major chord, or major triad (second audio clip in Figure 9). If we play instead the 1st, 2nd, 3rd, 5th, 8th, and 16th together, then it sounds like a more spacious version of the same chord — more like the grand, thunderous chord that opens the *Star Wars* music (third audio clip in Figure 9). If, on the other hand, we play the 6th, 7th, 9th, and 10th together, then we get what has famously been called the 'Tristan chord' (first audio clip in Figure 10):

Figure 10: Audio clips of the Tristan chord made up of the 6th, 7th, 9th, and 10th harmonic pitches shown in Figure 9. The first clip (QR code on the left, or zip file, etc., as noted in Figure 9) is the chord played with accurate harmonic-series tunings, as detailed in the caption to Figure 9. The second clip is the same chord in standard keyboard or equal-tempered tuning. If the first clip is played loudly, through a distorting audio system, then one usually hears not only the chord but also a low C corresponding to the 1st or 2nd harmonic. That's a consequence of the chord being accurately tuned to the harmonic series. The third clip (QR code on the right) sounds the chord in the spaced-out version used by Richard Wagner in the opening of *Tristan und Isolde* (see text), in equal-tempered tuning, along with the subsequent organic harmony changes that complete the opening phrase.

The first chord heard in Richard Wagner's opera *Tristan und Isolde* is a spaced-out version of the same chord, formed by moving the 7th and 9th harmonics down an octave and the 10th down by two octaves (third audio clip in Figure 10). Some people think that Wagner invented the chord, even though it was actually invented — I'd rather say discovered — much earlier. For instance, the chord occurs more than twenty times, in various spacings, in another famous piece of music, *Dido's Lament*, written by Henry Purcell about two centuries before *Tristan*.

The musical brain recognizes such chords as special even when slightly mistuned, as they always are when using standard keyboard or equal-tempered tuning. That's analogous to the visual brain recognizing curves such as those in Figure 7 as perfectly Platonic, even though they're slightly imperfect in reality. For instance, the Tristan chord sounds much the same if its tuning is changed from the pure-harmonic-series tuning, as heard in the first audio clip of Figure 10, to the standard keyboard tuning as heard in the second audio clip.

Claude Debussy — of *Clair de Lune* fame — seems to have been the first great composer to exploit the fact that *any* subset of pitches from a harmonic series is special to the musical brain. Notice that all the chords just mentioned are themselves harmonic-series subsets, when perfectly tuned, because each pitch has its own harmonic overtones and because whole numbers multiply together to give whole numbers. Toward the end of the nineteenth century, Debussy made extraordinary use of these insights, together with the organic-change principle for chord progressions, to open up what he called a new frontier in music.[89] Extending far beyond Wagner, Debussy's discoveries have been exploited across a range of twentieth-century genres, including vast amounts of film music and, for instance, the bebop jazz of Charlie Parker.

The organic-change principle for chord progressions involves small pitch changes, which as mentioned in Chapter 2 can be small in either of two different senses. These can now be stated more clearly. The first sense is the obvious one, closeness on the keyboard or guitar fingerboard. The second is the inverse distance between the notes in a harmonic series. So the 1st and 2nd harmonics are closest in this second sense. They are so close that musicians give them the same name, C or Doh in the case of Figure 9, even though they're far apart in the first sense, by a whole octave. The 2nd and 3rd are the next closest in the second sense, then the 3rd and 4th, and so on. This plus the need for invariant elements is almost all one has to know in order to master musical harmony, if one has a good ear — though admittedly the big textbooks on harmony and counterpoint are worth reading for their many useful illustrations, also showing the importance of how melodic lines or 'voices' move against one another. The organically-changing patterns can get quite complicated!

The musical brain is good at recognizing not just superposed melodic voices but also subsets from more than one harmonic series simultaneously, even when slightly mistuned and even when superposed in complicated combinations. Without this, we wouldn't have the sounds of *Star Wars* and symphonies and jazz bands and prog rock, and Stravinsky's ballet music *Petrushka,* for instance. Many powerful chords are made by superposing subsets from more than one harmonic series, to create what are sometimes called 'polychords'.

The chord known as the minor triad is a 3-note polychord in this sense, in its usual pure tuning. An example is given in the first audio clip of Figure 11. The figure caption describes how the chord is constructed from two different harmonic series. The first spooky chord in *Star Wars* is another 3-note polychord, called an augmented triad, as heard in the second and third audio clips of Figure 11 and again constructed from two different harmonic series, as described in the caption:

Figure 11: The first audio clip (QR code on the left, or zip file, etc., as noted in Figure 9) is an E minor triad in its usual pure tuning as a 3-note polychord. The lower two notes, E and G, are tuned as the 5th and 6th harmonics of the low C in Figure 9, while the upper two notes, G and B natural, are tuned as the 4th and 5th harmonics of the low G, 98.1Hz, that lies an octave below the 3rd harmonic

Caption continues on facing page

in Figure 9. Such polychordal tuning is often favoured by Western musical ears, not only in plain minor triads like this one but also within the Tristan chord, especially in spaced-out versions such as Wagner's or Purcell's. The second audio clip above is a 3-note polychord called an *augmented triad*, in this case made up of B flat, D, and F sharp, with the B flat and D tuned as the 4th and 5th harmonics of a low B flat, 116.54Hz, and the D and F sharp tuned as the 4th and 5th harmonics of a low D, 145.68Hz. The third audio clip is a fragment of the *Star Wars* music that uses the same augmented triad to make a spooky effect just after the grand opening. I should perhaps add that in the audio clips from the *Star Wars* music, I've transposed the music into the key of C major, slightly above its usual B flat major, to help with the comparisons.

Famously, Stravinsky's *Petrushka* features a strikingly powerful 6-note polychord made from two harmonic series. Musicians call it the Petrushka chord. It's a superposition of the major triads from the two series. They are the harmonic series in Figure 9, based on C, and the series based on an F sharp. Polychords of still greater complexity are often heard in 20th-century music, including film music. Many powerful film scores, such as those used in horror films, induce feelings of suspense and premonition through an ever-varying choice of harmonic-series subsets — superposed on each other within relentlessly complex patterns of organic change that don't settle down to anything simple or stable.

As detailed in the caption to Figure 9, there are differences between the harmonic-series pitches and the corresponding pitches on a standard keyboard. The differences are easily audible for the 7th and 11th harmonics and also audible, more subtly, for the 5th and 10th. Such differences have a significance beyond mere mistuning. They give us a valuable artistic resource. Singers, and players of non-keyboard instruments, in most genres, develop great skill in slightly varying the pitch from moment to moment as the music unfolds, navigating the slight tension between harmonic-series pitches and keyboard pitches for expressive purposes as with, for instance, 'blue notes' in pop vocals and in jazz. Blue notes often flirt with the pitch of the 7th harmonic, and sometimes with that of the 11th. A pianist can't sound a blue note, but a singer or saxophonist can, while the piano plays other notes.[90] (A further complication is that each piano note has slightly non-harmonic overtones.)

But — I hear you ask — *why* are these particular sets of pitches, such as those of Figure 9, so special to the brain, and what has all this to do with evolution and natural selection? Once again, the key word is simplicity. As already said, the defining property of a harmonic series is that its frequencies are whole-number multiples of the fundamental frequency or first harmonic. It follows that a sound wave created by superposing any set of pitches from a harmonic series takes a very simple form. The waveform precisely repeats itself at a single frequency. That's 65.4 times per second in the case of Figure 9. The Tristan chord, when tuned as in the first clip in Figure 10, produces just such a repeating waveform.

A famous mathematical theorem attributed to Joseph Fourier tells us that *any* repeating waveform corresponds to some superposition of pitches from a single harmonic series, as long as you allow an arbitrary number of harmonics. Repeating waveforms, then, are mathematically equivalent to sets or subsets of harmonic-series pitches — mathematically equivalent, even if psychologically very different.

Our neural circuitry is good at timing things. It has evolved to give special attention to repeating waveforms because they're important for survival in the natural world.

Many animal sounds are produced by vibrating elements in a larynx, or a syrinx in the case of birds. Such vibrations will often repeat themselves, to good accuracy, for many cycles, as the vibrating element oscillates back and forth like the reed of a saxophone or clarinet. So repeating waveforms at audio frequencies are important for survival because it can be important, for survival, to be able to pick out individual sound sources in a jungle full of animal sounds. This rather astonishing feat of perceptual model-fitting is similar to that of a musician skilled in picking out sounds from individual instruments, when an orchestra is playing. It depends on having a repertoire of internal model components that include repeating waveforms.

So exactly-repeating waveforms are among the simplest model components needed by the hearing brain to help identify sound sources, just as perfectly smooth curves are among the simplest model components needed by the visual brain to help identify objects. That's why exactly-repeating waveforms have the status of Platonic objects, or forms, for the hearing brain, just as perfectly smooth curves do for the visual brain. Both contribute to making sense of a complex visual scene, or of a complex auditory scene, as the case may be, while being as simple as possible.

In summary, then, for survival's sake the hearing brain has to be able to carry out auditory scene analysis, and therefore *has* to know about repeating waveforms — has to include them in its repertoire of unconscious model components — unconscious symbolic representations — available for fitting to the incoming acoustic signals in all their complexity. And that's mathematically equivalent to saying that the unconscious brain has to know about the harmonic series, and has to recognize it as special. Debussy's musical insights were indeed profound and far-reaching.

The accuracy with which our neural circuitry can measure the frequency of a repeating waveform reveals itself via musicians' pitch discrimination. Experience shows that the musical ear can judge pitch to accuracies of the order of a few cents, that is, to a few hundredths of a semitone, a few hundredths of the interval between adjacent pitches on a keyboard or guitar fingerboard. It used to be thought, incidentally, that our pitch discrimination is mediated by the inner ear's vibrating basilar membrane. That's wrong because, although the basilar membrane does carry out some frequency filtering, that filtering is far too crude to account for the accuracy of pitch discrimination.

Auditory scene analysis isn't exclusive to humans. So it should be no surprise to find that other creatures can perceive pitch to similar accuracies. The European cuckoo comes to mind. I've heard two versions of its eponymous two-note call in the English countryside. One of them matched the 6th and 5th harmonics with moderate accuracy, and the other the 5th and 4th. The composer Frederick Delius used both versions in his famous piece *On Hearing the First Cuckoo in Spring*. They're woven with exquisite subtlety into the gentle, lyrical music, starting around two minutes into the piece. Among other pieces of music quoting cuckoo calls the most famous, perhaps, are Beethoven's *Pastoral Symphony* and Saint-Saëns's *Carnival of the Animals*. Both use only the second version, matching the 5th and 4th harmonics.

In New Zealand, where I grew up, I heard even clearer examples of accurate avian pitch discrimination — much to my youthful astonishment. They came from a wonderfully feisty native bird called the tui, also called the parson bird because of its white bib worn against dark plumage, as shown in Figure 12:

Figure 12: New Zealand tui. Photograph courtesy of Keith Payne, by kind permission.

In some lighting conditions, such as those of Figure 12, a tui can look a bit like a European blackbird wearing a white bib. Tui have a large repertoire of complex and virtuosic calls, but as a schoolboy on summer holidays in the Southern Alps of New Zealand I encountered a tui that sang exceptionally simple tunes, using accurate harmonic-series pitches. The bird sounded those pitches with an accuracy well up to the standard of a skilled human musician, and distinctly better than the average cuckoo.

That was in the 1950s, before the days of cheap tape recorders, but I want to put on record what it sounded like. I well recall two of the tunes. Imagine a miniature xylophone with a wonderfully clear musical sound, echoing through the trees of a beech forest in the Southern Alps. Figure 13 shows a transcription of one of the tunes along with an audio clip reconstructing it:

Figure 13: First tui song. The QR code points to a corresponding audio clip. As before, in some e-readers you can also click on the QR code; and in case none of this works, the clip is available in the zip file mentioned earlier, from https://www.worldscientific.com/worldscibooks/10.1142/13429#t=suppl.

It uses the 4th, 5th, and 10th harmonics from Figure 9, though two octaves higher. The sounds were short like those of a real xylophone, but clearer and more resonant. Figure 14 shows the other tune:

Figure 14: As in Figure 13 for the second tui song.

It uses the 5th, 6th, 8th, and 10th together with two notes from another harmonic series. To a human musician, both tunes are in the key of C major. The bird *always* sang in C major, sounding the notes as accurately as any human musician. In the second tune, the third and fourth notes are the 6th and 5th harmonics of B flat. The rhythms were exceptionally simple and regular, almost exactly as notated in the figures. Each tune ended abruptly with a complicated, glittering short burst of sound, denoted by a cross, which I could imitate only crudely in the reconstructions. Such complicated sounds, with no definite, single pitch, are more typical of tui utterances. The bird sang just one tune, or slight variants of it, in the summer of one year, and the other tune in another year, sometime in the 1950s.

Even though tui are famous for their skills in mimicry, I think we can discount the possibility that this bird learnt its tunes by listening to a human musician. The bird's territory was miles away from the nearest human habitation, other than our holiday camp; and we had no radio or musical instrument apart from a guitar and a descant recorder. The recorder is a small blockflute with a pitch range overlapping the bird's. When the bird was around, I used to play various tunes on the recorder, in the same key, C major, but never got any response. It was as if the bird felt my efforts to be unworthy of notice. It usually fell silent, then later on started up with its own tune again.

An internet search turns up many recordings of tui song, but I have yet to find an example remotely as simple and tuneful as the C major songs I heard. And my impression at the time was that such songs are exceptional among tui. I did, however, find a more complex tui song that again demonstrates accurate pitch discrimination. It's interesting in another way because one hears two notes at once, accurately tuned against each other. As is well known, the avian syrinx can be used like a pair of larynxes, to sound two notes at once. Figure 15 shows a transcription:

Figure 15: Transcription of a more complex tui song, with audio clips of it courtesy of Les McPherson, by kind permission. The first audio clip (QR code on the left, or zip file etc., as noted in Figure 13) is from McPherson's original recording. It contains two partial and two complete repetitions of the song. Tui often repeat fragments of songs. The second audio clip is the song slowed to half speed, to make the details more accessible to human hearing.

At the first occurrence of two notes together, they're accurately tuned to the spacing of the 3rd and 4th harmonics (of B natural), the interval that musicians call a perfect fourth, notoriously sensitive to mistuning. To my ear, when I listen to the half-speed version (second audio clip in Figure 15), the bird hits the perfect fourth very accurately before sliding up to the next smallest harmonic-series spacing, that of the 4th and 5th harmonics (of G), what's called a major third, again tuned very accurately. At half speed this musical fragment is playable on the violin, complete with the upward slide, as I've occasionally done in lectures.

Another New Zealand bird that's known to sing simple, accurately pitched tunes is the North Island kokako, a crow-sized near-ground dweller. Figure 16 presents an example:

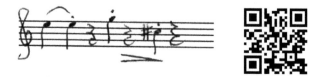

Figure 16: North Island kokako song. The transcription corresponds only to the start of the song recorded in the audio clip, again courtesy of Les McPherson by kind permission (QR code or zip file etc., as noted in Figure 13). The song continues in a more complicated way ending with the 8th harmonic of 110Hz A, to my ear creating a clear key-sense of A major. The first three notes, those shown in the transcription, are close to the 6th, 7th, and 5th harmonics of the same A. Listen carefully! The second note is an avian 'blue note'.

I want to mention one more connection between music and mathematics. It's another connection not mentioned in the standard accounts, confined as they are to games with numbers. Of course, composers have always played with numbers to get ideas, but that's beside the point. The point here is that there are musical counterparts to illusory contours like that in Figure 8. Music has its own calculus of variations! Listen to the first audio clip in Figure 17:

Figure 17: Audio clips from the opening of Mozart's piano sonata K 545, again available, alternatively, from the zip file etc. noted in Figure 13.

It's the opening of Mozart's piano sonata K 545, whose second movement was quoted in Figure 5. After the first eight seconds or so, one hears a smooth, flowing passage of fast notes that convey a sense of continuous motion, a kind of musical smooth curve, bending upward and downward. Mozart himself used to remark on this smoothness. In his famous letters he would describe such passages as flowing like oil when played well enough. But, as with the black segments in Figure 8, there's no smoothness in the actual sounds. The actual sounds are abrupt and percussive.

Of course, hearing doesn't work in the same way as vision. The analogy is imperfect. To give the impression of smoothness, in the musical case, the adjacent notes need to have similar loudness and to be spaced evenly in time. Mozart admitted that he'd had to practise hard to get the music flowing like oil. That's an example of what I meant by the fine control of rhythm 'to a few milliseconds or thereabouts', by world-class musicians.

When the notes are not spaced evenly in time, as in the second audio clip in Figure 17, the smoothness — Mozart's 'oiliness' — disappears. That's perhaps reminiscent of the *outer* ends of the black segments in Figure 8.

Coming back to musical pitch perception for a moment, if you're interested in perceived pitch then you may have wondered how it is that violinists, saxophonists, singers, and others can use the kind of frequency variation called 'vibrato' while maintaining a clear and stable sense of pitch.

Vibrato can be shaped to serve many expressive purposes and is an important part of performance in most Western and other musical genres. The performer modulates the frequency by amounts far greater than the pitch discrimination threshold of a few cents, up and down over a range that can be as much as a hundred cents — a semitone — and sometimes even more. The timing of the frequency modulations can vary within a range of about 4 to 7 complete cycles per second, depending on the style and expressive purpose. There are no corresponding fluctuations in perceived pitch. That, however, depends on the fluctuations being fast enough.

A recording played back at half speed or less tends to elicit surprise when heard for the first time. One then hears a gross wobble in the pitch. In the first audio clip of Figure 18, lasting about 26 seconds, a violin playing alone begins a quiet little fugue before a piano joins in. The use of vibrato is rather restrained. Yet, when played at half speed, as in the second clip, the pitch-wobble becomes surprisingly gross and unpleasant:

Figure 18: Audio clips illustrating aspects of violin vibrato. As before, if the QR codes do not work, the clips are available, alternatively, from the zip file etc. noted in Figure 13 and earlier. The first clip presents the opening of the little fugue at normal speed. The second is the opening played at half speed, showing how a vibrato — even a restrained vibrato — then turns into what, to my ear at least, is an ugly pitch-wobble. The third clip presents the whole piece[91] at normal speed in case you, dear reader, are curious to hear how it continues.

It seems that in order to judge pitch the musical brain does not carry out Fourier analysis of the incoming data but, rather, carries out a more flexible kind of model-fitting — in effect counting waveform cycles over timespans up to two hundred milliseconds or thereabouts, long enough to span a vibrato cycle. This idea was called the 'long-pattern hypothesis' by Paul Boomsliter and Warren Creel, in a classic discussion based on some ingenious psychophysical experiments.[92] It accounts not only for the pitch-stability of vibrato but also for several other phenomena familiar to musicians, including the tolerance to slight mistuning of chords already mentioned, as with the Tristan chord of Figure 10, second audio clip.

Another musically significant aspect of vibrato is that it can powerfully influence the perceived tone quality. For instance, quality can be perceived as greatly enriched when the strengths of different overtones fluctuate out of step with each other, as happens with violins and other bowed-stringed instruments.[93] It seems that the unconscious brain has a special interest in waveforms that almost repeat themselves but vary their shapes, as well as their periods, continuously over a long-pattern timespan. That's another example of organic change.

Stepping back from all those details, I realize that I haven't yet fully answered my opening question, "what is music?"

To be sure, part of the answer is that, as with mathematics, as already suggested, music is something that comes from our unconscious interest in abstract, organically-changing patterns and our love of playing with such patterns, going all the way back to juvenile play, "that deadly serious rehearsal for real life". Playing with abstract patterns is central to brain development.

What I've hardly touched on, though, is the further significance of music as, evidently, part of our emotional and social intelligence promoting group solidarity. As such it could well be even more ancient than language — yet another consequence of the multi-timescale co-evolution of genomes and cultures, stretching back in time to ancestral primates that, perhaps, made use of music in family bonding like today's gibbons. Such bonding and its music might then have fed into proto-language.

Be that as it may, our own response to music quite evidently goes back to mothers' singing and baby talk. For us, music does overlap with language, very intimately.[14] And for older children, and adults, myself included, musical experience can be intense and can evoke a feeling of being part of something greater than oneself. For some individuals, musical experience is almost the *same* as religious experience.[39] The Scottish composer James MacMillan testifies to this. And the powerful emotional effects of music in crowds can, on the other hand, be part of our versatility and flexibility in adopting group identities. These collective social effects further illustrate the point made so clearly by the work of Stephen Reicher and Clifford Stott,[73] that human group identities needn't be rigid, filter-bubbled, and polarized but can instead be flexible, versatile, multifarious, circumstance-dependent, and adaptable.

* * *

Let's return for just a moment to theoretical-physics fundamentals. Regarding models that are made of mathematical equations, there's an essay that every physicist knows of, I think, by the famous physicist Eugene Wigner, about the "unreasonable effectiveness of mathematics" in representing the outside world. But what's unreasonable is not the fact that mathematics comes in. As I keep saying, mathematics is just a means of handling many possibilities at once, in a precise and self-consistent way, and of appreciating the associated abstract patterns and their surprising interrelations.

What's unreasonable is that *very simple* mathematics comes in when you build and test accurate models of, for instance, the subatomic world. It's not the mathematics that's unreasonable; it's the simplicity. So I'd prefer to talk about the unreasonable simplicity of, for instance, the subatomic world.

It just happens that, at the level of electrons and other subatomic particles, things look astonishingly simple. That's just the way the world seems to be, at that level. And of course it means that the corresponding mathematics is simple, too, at that level. One of the greatest *unanswered* questions in physics is whether things stay simple, or not, when we zoom in to the lower levels and the far smaller 'Planck lengthscale' at which quantum phenomena and gravity mesh together, the levels that would need to be included in a 'Theory of Everything'.

As is well known, we're now talking about lengthscales roughly of the order of a hundred billion billion times smaller than the diameter of a proton, and ten million billion billion times smaller than the diameter of a hydrogen atom — well beyond the range accessible to observation and experimentation. At those small scales, things might for instance be complex and chaotic, like turbulent fluid flow, with order emerging out of chaos and making things look simple only at much larger scales.

Such possibilities have been suggested, for instance, by my colleague Tim Palmer, who has thought deeply about these issues — and about their relation to the vexed questions at the foundations of quantum theory[94] — alongside his better-known work on the chaotic dynamics of weather and climate.

So let's turn now, at last, to the climate problem, by far the most complex of all the problems that confront us, a vast jigsaw of interacting pieces, from global-scale atmospheric and oceanic circulations all the way down to the microscopic scales of bacteria, viruses, and molecules. What's at stake is existential for human civilization even though, I'll argue, probably not existential for life on Earth. I've written the rather long Postlude that follows in the hope of clarifying what science can and can't tell us, today, about the problem. I can't entirely avoid the politics, but will nevertheless try to keep the main focus on our progress toward in-depth scientific understanding.

Postlude

The amplifier metaphor for climate

Journalist to scientist regarding a windstorm, firestorm, blizzard, flood, or other weather extreme: "Tell me, Professor So-and-So, is this a one-off extreme, pure chance, or is it due to climate change?"

Well — once again — dichotomization makes us stupid. The professor needs to say "This isn't an either-or. It's both. Climate change produces long-term upward trends in the *probabilities* of extreme weather events, and in their peak intensities." As the extremes become more intense, more frequent, and more devastating, these points have at long last gained traction and indeed — as I go to press with this second edition — have begun to be noticeable in mainstream politics even if not, as yet, at high enough priority.

Politics aside, though, my main point in this Postlude will be that the worst scientific uncertainties about future climate change concern the possibility of tipping points. There might or might not be a domino-like succession, or cascade, of tipping points that might or might not, after an uncertain number of centuries, take us into a hothouse climate like that of the early Eocene, around 56 million years ago, with sea levels up by around 70 metres and extremes of storminess beyond human experience. Such worst-case scenarios might or might not be reversible by future technologies. And they are outside the scope of today's climate prediction models. We have no way to rule them out with complete confidence. So there has never in human history been a stronger case for applying the precautionary principle. Today there is no room for doubt — even from a purely financial perspective — about the need to reduce net greenhouse-gas emissions urgently and drastically, far more than is possible through so-called 'offsetting'.

* * *

Here's one way to get into what's involved scientifically. Chapter 2 mentioned audio amplifiers and two different questions one might ask about them: first, what powers them, and second, what they're sensitive to. For the climate system, switching off the power corresponds to switching off the Sun. But in that system is there anything corresponding to an amplifier's sensitive input circuitry?

For many years now, we've had a clear answer: yes. The climate system can be thought of as a powerful amplifier whose sensitive inputs include greenhouse-gas emissions from human activities. And it's responding to those inputs by becoming more active, with larger fluctuations, including greater weather extremes. Today that's a practical certainty. Larger fluctuations generally means larger in both directions — hot and cold for instance, wet and dry. What's uncertain is how fast the system will respond in future, how far it will go, and by precisely what stages including possible tipping points. Such questions are, as just suggested, outside the scope of today's climate prediction models. Despite many recent improvements, the models are far from representing the full complexity of the system. And it so happens that they've underestimated the response so far — for reasons I'll come to.

Today's understanding of all this has come from looking at past as well as present climates and at all the inputs, human and non-human. Among the amplifier's non-human inputs are volcanic eruptions, small changes in the Sun's power output, and, more importantly, small orbital changes — small changes in the Earth's tilt and in its orbit around the Sun. These last can be regarded as small inputs because they hardly change the total power delivered by the Sun, but do slightly change its distribution over the Earth's surface, in ways that I'll discuss. During the past 400 millennia or so the response to the small orbital changes was a sequence of large climate changes, the last four glacial–interglacial cycles illustrated by the ice-core data in Figure 3, 'glacial cycles' for brevity. And 'large' is a bit of an understatement. For instance, as the ice sheets changed, some of the associated sea-level changes were well over a hundred metres, as noted in Chapter 3. That's huge by comparison with anything projected for the current century.

The main human input today is the injection, or emission, of carbon dioxide into the atmosphere. It comes from a variety of activities but most of all from burning fossil fuels such as coal, oil, and natural gas. Carbon dioxide, whether injected naturally or artificially, has a central role in the climate-system amplifier not only as a plant nutrient but also as our atmosphere's most important *non-condensing greenhouse gas*.

That central role is crucial to climate behaviour in general, and to the huge magnitudes of the last four glacial cycles in particular. Those cycles depended not only on small orbital changes but also on natural injections of carbon dioxide into the atmosphere from the deep oceans, as we'll see.

Of course, to think of such natural injections as 'inputs' is strictly speaking incorrect, except as a thought-experiment, but they're part of the amplifier's sensitive input circuitry. The ice-sheet dynamics will also prove to be sensitive.

The physical and chemical properties of so-called greenhouse gases are well established and uncontentious, with very many cross-checks. Greenhouse gases in the atmosphere make the Earth's surface roughly 30°C warmer than it would otherwise be.[95] For reasons connected with the properties of heat radiation, any gas whose molecules have three or more atoms can act as a greenhouse gas. More precisely, to interact strongly with heat radiation the gas molecules must have a structure that supports a fluctuating electric 'dipole moment' at the frequency of the heat radiation, of the order of tens of trillions of cycles per second. Examples include not only carbon dioxide and water vapour, each with three atoms per molecule, but also for instance nitrous oxide, with three atoms, and methane, as in natural gas, with five atoms. By contrast, the oxygen and nitrogen molecules making up the bulk of the atmosphere have only two atoms, and are nearly transparent to heat radiation. Ref. 95 gives an authoritative discussion.

One reason for the special importance of carbon dioxide is its great chemical stability as a gas. Other natural carbon-containing, non-condensing greenhouse gases such as methane tend to be converted fairly quickly into carbon dioxide. Fairly quickly means within a decade or two, for methane. And of all the non-condensing greenhouse gases, carbon dioxide has always had the most important long-term warming effect, not only today but also during the glacial cycles. That's clear from its chemical stability and from the ice-core data, to be discussed below, along with the well established heat-radiation physics, all cross-checked by very many accurate measurements.

The role of water vapour is also central, but entirely different. It too is chemically stable and has great importance as a greenhouse gas. But unlike carbon dioxide it can and does condense or freeze, in vast amounts, for instance as cloud, rain, and snow, while copiously resupplied by evaporation from the tropical oceans and elsewhere. This solar-powered supply of water vapour — sometimes called 'weather fuel' because of the latent heat energy put in by the Sun, and released on condensing or freezing — dwarfs any human input and makes it part of the climate-system amplifier's power-supply and power-output circuitry, rather than its sensitive input circuitry.

Air can hold around six or seven percent more weather fuel for every degree Celsius rise in temperature. This comes from a robust and well established piece of physics called the Clausius–Clapeyron relation. So global warming is also *global fuelling*.

Some of the amplifier's power output drawing on weather fuel takes the form of tropical and extratropical thunderstorms and cyclonic storms, including those that produce the most extreme rainfall, flooding, and wind damage. The latent energy released dwarfs the energies of thermonuclear bombs. Cyclone Idai, which caused widespread devastation in Mozambique, and Hurricane Dorian, which flattened large areas of the Bahamas, and the typhoons impacting the Philippines — and many other examples — remind us of what those huge energies mean in reality.

It's reasonable to expect that more weather fuel will make extremes more extreme. An especially clear example is that of thunderstorms. A thunderstorm is like a giant vacuum cleaner, powered by the weather fuel it pulls in from its low-level surroundings. So, if it's surrounded by air containing more weather fuel — other things being equal — it's more vigorous and pulls the fuel in faster. That's a very robust positive feedback, a robust self-reinforcing process, in the runup to peak intensity. The peak intensity is greater and comes sooner. The consequences can include flash flooding, and damage by lightning strikes. Such peak intensities are completely missed by the climate prediction models, whose spatial resolution is far too coarse to describe the airflow into thunderstorms.

Please note that I'm talking about the strongest and most devastating thunderstorms, not the average thunderstorm. That point needs to be remembered when looking at statistical summaries of data, which tend to focus on averages and to hide the extremes, even if they're present in the data.

I'll postpone the discussion of other examples, since they're not so simply and robustly arguable. They depend on a more complex web of cause and effect. The same points apply, though. The most extreme behaviours, not only of thunderstorms but also of cyclones and other weather systems, tend to get hidden in statistics and in any case are outside the scope of the climate prediction models. That's again because the most extreme behaviours usually, in one way or another, involve small spatial scales that the models cannot resolve. I'm not, by the way, saying that the models are useless. In many ways they're a crucially important part of our hypothesis-testing toolkit, when used within their limitations.

A further point is that the response of the system to carbon-dioxide emissions, or injections, is practically speaking irreversible[96] thanks to properties of the so-called 'carbon cycle', to be discussed below. And it's clear that if we carry on as now, continuing to subsidize fossil fuels[97] and to build new coal mines, oil wells, and gas wells, then the resulting changes will not only be irreversible but also very large indeed. They're already more than large enough today, and expensive enough, as the catastrophic Pakistan floods of 2022 have reminded us. As said earlier the main uncertainties are not about the changes being large but only about just how large, just how far they'll go, and by precisely what stages — perhaps even as a cascade of tipping points taking us into runaway climate change, toward something like the storminess of the early Eocene.

In climate science the term 'runaway' has more than one meaning, along with other technical terms such as 'the' climate sensitivity to carbon dioxide. Human language is, indeed, a conceptual minefield. I'll return to these issues and to the question of tipping points, mentioning some of the possible mechanisms. As for storminess in the early Eocene, we'll see that there's strong evidence for it, consistent with the positive-feedback argument about thunderstorms and flash flooding. Part of that evidence is the existence today of whales and dolphins.

In what follows, then, I'll try to explain in more detail what science can and can't tell us about climate change as clearly, simply, accessibly, and dispassionately as I can — along with the implications under various assumptions about the politics.

Is such an exercise useful at all? The optimist in me says it is. And I hope that you might agree because, after all, we're talking about the Earth's life-support system and the possibilities for some kind of future civilization. Life on Earth will almost certainly survive — even if there's runaway into a new Eocene — but human civilization might not.

To build understanding it's useful, as already suggested, to look at past climates. In doing so I'll draw on the wonderfully meticulous work of many scientific colleagues including the late Nick Shackleton, and his predecessors and successors, who have extracted large amounts of information about past climates from the geological record. They're among my scientific heroes.

Past climates are our main source of information about the workings of the real system, taking full account of its vast complexity all the way down to the details of cyclones, thunderstorms, forest canopies, soil ecologies and mycorrhizal 'wood wide web' networks, ocean plankton (see below), archaea, bacteria, and viruses, and the tiniest of turbulent eddies. And in case you think tiny eddies can't be important I should point out that they've long been known, from the classic work of Walter Munk,

Chris Garrett, Carl Wunsch, and others, to be crucial to the large-scale temperature structure of the oceans all the way down from the surface to the deepest, coldest 'abyssal' waters. That structure is shaped by the vertical mixing due to millimetre-scale eddy motion and is crucial, in turn, to the way the deep oceans store and release carbon.

* * *

In recent decades there's been a powerful and well-funded series of disinformation campaigns on climate, aimed at promoting fossil-fuel burning and its continued subsidization. Scientists' efforts to deepen our understanding of the climate problem, in all its fearsome complexity — daring to look at it from more than one angle — have been portrayed as duplicitous and dishonest. The 'climategate' campaign of 2009 was an example. The disinformers used out-of-context quotes from private emails between scientists who were trying to estimate atmospheric temperatures over the past millennium. Such estimates have formidable technical difficulties, such as allowing for the effects of pollutants on tree rings. The scientists were not being dishonest; they were struggling to work out how to allow for such effects. For me it's a case of *déjà vu*, because the earlier ozone disinformation campaigns — which I encountered at close quarters during my own professional work — were strikingly similar.

It's been known for some years now that the similarity was no accident. According to extensive documentation discussed in Ref. 24 — including formerly secret documents exposed through anti-tobacco litigation — the climate disinformation campaigns were seeded, originally, by the same professional disinformers who masterminded the ozone campaigns and, before that, the tobacco companies' lung-cancer campaigns. The secret documents describe how to manipulate the newsmedia, pressing binary buttons and spreading confusion in place of understanding.

For climate the confusion spread into a number of scientific communities including some influential physicists who weren't, to my knowledge, among the *professional* disinformers and their sponsors but who tended to focus too narrowly on the shortcomings of the climate prediction models, ignoring the many other lines of evidence. And the disinformation campaigns and their political fallout are, of course, threats to other branches of science as well, and indeed to the very foundations of good science. The more intense the politicization, the harder it becomes to live up to the scientific ideal and ethic.

One reason why the amplifier metaphor is useful is that it helps to expose one of the climate disinformers' favourite tactics. They use the copious supply of water vapour from the tropical oceans to suggest that the relatively small amounts of carbon dioxide in the atmosphere are unimportant for climate. That's like focusing on the amplifier's power-output circuitry and ignoring the input circuitry, exactly the 'energy budget' mindset mentioned in Chapter 2, conflating 'small' with 'unimportant'. Again, I get a sense of *déjà vu*. The disinformers used the same tactic with the ozone hole, saying that the pollutants alleged to cause it were present in such tiny amounts that they couldn't possibly be important. Well, for stratospheric ozone there's an amplifier mechanism too. It's called chemical catalysis. A single molecule of pollutant destroys thousands of ozone molecules.

A point to note about the prediction models, for ozone as well as for climate, is that their shortcomings have tended, as said earlier, to make them err on the side of underprediction rather than overprediction. In both cases the real effects of the pollutants have turned out to be worse than predicted. That's the exact opposite of what the disinformers have always claimed about the models, namely that the models are bad (true, especially of the earliest versions, both for ozone and for climate) and 'therefore' that they overpredict (false). And on top of the climate models' *under*prediction — especially as regards the probabilities of weather extremes and, I'll argue, the rate of sea-level rise — scientific caution and fear of being accused of scaremongering has given rise to a tendency, among climate scientists, to further understate the risks.

In all humility, I think I can fairly claim to be qualified as a dispassionate observer of the climate-science scene. My own professional work was never funded for climate science as such. However, my professional work on, for instance, the ozone hole and the fluid dynamics of the great jet streams has brought me close to research issues in the climate-science community.

Those of its members whom I know personally include many brilliant thinkers and innovators. They respect the scientific ideal and ethic. They try hard to cope with daunting complexity. Again and again, I've heard members of the community giving careful conference talks on the latest findings, and taking questions. They're concerned about the shortcomings of models, about the difficulty of weeding out data errors, and about the need to avoid superficial viewpoints and exaggerated claims. Those concerns are reflected in the restrained and cautious tone of the vast reports published by the Intergovernmental Panel on Climate Change (IPCC). The reports make heavy reading but contain reliable technical information about the basic physics and chemistry I'm talking about, such as the contributions from various greenhouse gases and the magnitude of greenhouse-gas warming as compared with variation in the Sun's output — the slightly variable solar 'constant'. Another source of reliable technical information is Ref. 95, the big textbook by my colleague Raymond Pierrehumbert.

As it happens, my own professional work has included research on solar physics. My judgement on that aspect is that recent IPCC assessments of solar variation are substantially correct, namely that solar variation is too small to compete with the ongoing carbon-dioxide injections from human activities. That's based on recent improvements in our understanding of solar physics, to be referred to below. I mention them here, though, because another of the climate disinformers' tactics has been to claim the opposite, that solar variation dominates.

* * *

Let's pause for a moment to draw breath. I want to get on with explaining how past climates have informed us about these issues, using the latest advances in our understanding. I'll discuss the leading implications and the reasoning behind them, and the different senses of the technical terms 'runaway' and 'climate sensitivity'. I'll address the issue of sea-level rise. The focus will be on what we can say independently of fine details within the climate system, and independently of the shortcomings of the climate prediction models.

* * *

The first point to note is that human activities have been increasing the carbon dioxide in the atmosphere by amounts that, regarded as amplifier input, are already large.

A natural benchmark against which to measure such increases is the *observed range of variation* of atmospheric carbon dioxide during the past several hundred millennia, when the climate system was close to its present state — far closer than in the early Eocene, for instance. The range of variation is accurately determined by the Antarctic ice-core data. That's one of the hardest, clearest, most unequivocal pieces of evidence we have. It comes from the ability of ice to trap air, beginning with compacted snowfall, giving us clean air samples from which carbon-dioxide concentrations can be reliably measured, exploiting the chemical stability of carbon dioxide as a gas.

Typical results include those shown in the left-hand half of Figure 3, in the bottom graph. In round numbers the natural range of variation of atmospheric carbon dioxide, across the last four glacial cycles, is of the order of 100 ppmv, 100 parts per million by volume. Atmospheric carbon dioxide increased by amounts of this order each time the system underwent a so-called 'deglaciation' — a transition from the coldest to the warmest conditions.

The increase since pre-industrial times now exceeds 120 ppmv. In round numbers, we've gone from a glacial 180 ppmv through a pre-industrial 280 ppmv up to today's values, well over 400 ppmv, not shown in Figure 3. So we've already had a total, natural plus artificial, carbon-dioxide injection producing a change that exceeds twice the natural range of variation. In the fossil-fuel-driven 'business as usual' scenario, as it's called, values increase to around 800 ppmv by the end of this century. An increase from 180 to 800 ppmv is an increase of the order of *six times* the natural range of variation. Whatever happens, therefore, the climate system will be like a sensitive amplifier subject to a large new input signal, the only question being just how large.

Today, with atmospheric carbon dioxide well over 400 ppmv, the climate response is already substantial, along with its weather extremes. That's despite what's technically called the logarithmic behaviour of the greenhouse warming effect (e.g. Sec. 4.4.2 of Ref. 95), implying that the magnitude of the warming effect is described by an upward-sloping graph whose slope decreases as atmospheric carbon dioxide increases. Some climate disinformers have claimed, incorrectly, that that in itself makes the climate response negligible.

Recent work[98-104] has been clarifying what happened during the deglaciations and, in particular, how the 100 ppmv increases in atmospheric carbon dioxide came about, and with what effect. The increases were due mainly to natural injections of carbon dioxide into the atmosphere from the deep oceans and were part of a complex interplay of ice-sheet, sea-ice, and ocean-circulation changes. Those changes were initiated by small orbital changes that favoured summertime melting on the northern ice sheets. Crucially, as we'll see, the melting was then increased by a positive feedback from the greenhouse warming due to the natural carbon-dioxide injections.

The deglaciations illustrate how sensitive the climate-system amplifier can be. What I'm calling its sensitive input circuitry includes not only ice-sheet dynamics but also what's called the natural 'carbon cycle'. That so-called 'cycle' refers to the way in which the more active forms of carbon — those most directly relevant to climate — are moved around within the Earth system. The cycle depends on deep-ocean storage of carbon dioxide, mostly as bicarbonate ions.[95, 100] It depends on chemical and biochemical transformations on land and in the oceans, and on complex groundwater, atmospheric, and oceanic flows down to the smallest turbulent eddies, on sea-ice cover, upper-ocean layering, and plankton populations, including the plant-like *phytoplankton* that feed on sunlight in the upper ocean and pull carbon dioxide out of the atmosphere. It depends on the dynamics of biological and ecological adaptation and evolution. Nearly all of this is outside the scope of the climate prediction models. Much of it is also outside the scope of specialist carbon-cycle models,[96] if only because such models grossly oversimplify the transports of carbon and other plankton nutrients by fluid flows, within and across the layers of the sunlit upper ocean, for instance.

How much deep-ocean carbon is stored or released depends on a delicate competition between storage rates and leakage rates. For instance, there's storage via plankton corpses and excrement sinking from the sunlit upper ocean into the deepest, abyssal waters. That storage process is strongly influenced, it's now clear, by details of the ocean circulation, especially the highly variable Atlantic 'overturning circulation', linking the Greenland area to Antarctica and the circumpolar Southern Ocean.[70] The overturning circulation has deep and shallow branches and has strong effects on interhemispheric heat transport, on heat and gas exchange with the atmosphere, and on phytoplankton nutrient supply. All this has been under intense scrutiny in recent research.[98–103]

On the interplay of events during deglaciations, the records showing the greatest detail are those covering the last deglaciation. They include records in ice cores, in caves, and in the sediments under lakes and oceans.[70] Around 18 millennia ago, after the onset of an initiating orbital change, atmospheric carbon dioxide started to build up well beyond a glacial minimum of 185 ppmv toward the pre-industrial 280 ppmv. Around 11 millennia ago, it was already close to 265 ppmv. More details are given in Figure 20 below.

The 80 ppmv increase, from 185 to 265 ppmv, was the main part of what I'm calling a natural injection of carbon dioxide into the atmosphere. It must have come from deep within the oceans since, in the absence of injections by humans, it's only the deep oceans that can store the required amounts of carbon. Indeed, storage on land worked mostly in the opposite direction as ice retreated and forests spread, pulling some of the injected carbon dioxide back out of the atmosphere.

As for what happened to sea levels during the deglaciation, the evidence is unequivocal. There are two independent lines of evidence, a point to which I'll return. The fastest rate of sea level rise was between about 16 and 7 millennia ago. The total sea level rise over the whole deglaciation was well over 100 metres. The data in

Figure 20 show 140 metres. It required the melting of huge volumes of land-based ice. Exactly how the ice melted is one of the greatest unsolved problems in climate science. Our ability to model ice flow is very limited. That's another point to which I'll return because it bears directly and urgently on today's problems.

As for the orbital changes, they're well known and have been calculated very precisely and very securely over far greater timespans. Such calculations, using a suitable version of Newton's equations, are possible because of the remarkable stability of our solar system's planetary motions. The orbital changes include a small oscillation in the tilt of the Earth's rotation axis — between angles of about 22° and 24°, repeating every 40 millennia or so — and an axial precession, which keeps reorienting the axis relative to the stars.

The orbital changes redistribute solar heating in latitude and time, while hardly changing its average over the globe and over seasons — more precisely the average *insolation*, the incident solar power per unit area. Figure 19 illustrates one of the resulting effects, the variation of midsummer insolation at latitude 65°N. Time in millennia runs from right to left:

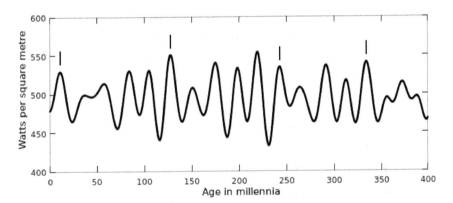

Figure 19: Midsummer diurnally averaged top-of-atmosphere insolation at 65°N, in watts per square metre, from calculations by André Berger and co-workers. They assume constant solar output but allow for orbital changes in the manner pioneered by Milutin Milanković. Data kindly supplied by Dr Luke Skinner.

The succession of peaks results from the axial precession in combination with other, slower changes, including what's called the 'apsidal precession' of the Earth's orbit, and slow changes in the eccentricity of the orbit. One gets a peak when closest to the Sun with the North Pole tilted toward the Sun. Such a peak boosts summertime melting on the northern ice sheets.

However, such melting is not in itself enough to trigger a full deglaciation. Only one peak in every five or so is associated with a full deglaciation. They're the peaks marked with vertical bars. The timings can be checked from Figure 3. The marked peaks were accompanied by the biggest carbon-dioxide injections. It's noteworthy that, of the two peaks at around 220 and 240 millennia ago, it's the smaller peak around 240 millennia that's associated with the bigger carbon-dioxide and temperature response in Figure 3. The bigger peak around 220 millennia is associated with a *smaller* response.

In terms of the amplifier metaphor, therefore, we have an input circuit whose sensitivity varies with time. In particular, the sensitivity to high-latitude insolation must have been greater around 240 than around 220 millennia ago. That's another thing we can say independently of the climate prediction models.

There are well-known reasons to expect such variations in sensitivity. One is that the system became more sensitive when it was fully primed for the next big carbon-dioxide injection. To become fully primed it needed to store enough carbon in the deep oceans. Storage was favoured in the coldest conditions, which tended to prevail during the millennia preceding full deglaciations. How this came about is now understood in outline, with changes in ocean circulation playing a key role alongside phytoplankton fertilization by mineral nutrients in airborne dust.[103] The ice-core records show that more dust was blowing around in the coldest conditions,[70] increasing the fertilization and hence the storage rate.

Also important was a different priming mechanism, the slow buildup and areal expansion of the great northern land-based ice sheets. The ice sheets slowly became more vulnerable to melting in two ways, first by expanding equatorward into warmer latitudes, and second by pushing the Earth's crust downward, gradually taking the upper surface of the ice down to warmer *altitudes*. This ice-sheet-mediated priming mechanism made the system still more sensitive. Specialized model studies support the view that both priming mechanisms were important precursors to a full deglaciation.[104]

As already suggested, however, our ability to model ice flow and melting is very limited, adding to our uncertainties about the near future as well. There's strong evidence today that parts of the Greenland ice sheet are now melting at increasing rates, as well as parts of the Antarctic ice sheet. Of special concern today is the Thwaites Glacier area in West Antarctica, where increasingly warm seawater is intruding sideways underneath the ice. In that area and elsewhere in West Antarctica, the ice is, or was, grounded below sea level.

Ice-flow modelling is peculiarly difficult because of the complex, highly variable fracture patterns and friction in glacier-like 'ice streams' within the ice sheets, over areas whose sizes, shapes, and frictional properties are hard to predict. They involve complex 'hydrological networks' of subglacial flow fed, for instance, by surface meltwater chiselling its way down through the ice by 'hydrofracture', whereby crevasses are forced to expand downward.[105, 106] The resulting basal lubrication can greatly increase bulk ice-flow rates.

One of the many challenges to ice-flow modelling is that posed by the Heinrich events, as they're called, that are marked by layers of rocky debris found in Atlantic ocean sediment cores. Massive ice flows occurred intermittently during some of the colder times within the past several hundred millennia, depositing large amounts of glacial debris into icebergs and thence into ocean sediments. The deposits show that the ice-flow dynamics occasionally became sensitive — during colder rather than warmer times — possibly through geothermal heating and the formation of huge ice streams. But the details are still largely unknown. Any claim to have created a comprehensive ice-flow model would need to show that it can explain Heinrich events.

As regards the deglaciations and the roles of the abovementioned priming mechanisms — ice-sheet buildup and deep-ocean carbon storage — two separate questions must be distinguished. One concerns the huge *magnitudes* of the last five full deglaciations. The other concerns their *timings*, roughly every 100 millennia.

It's hard to assess the timescale for ocean priming because here, too, our modelling ability is very limited, not least regarding the details of sunlit upper-ocean circulation and layering, where the phytoplankton live and feed on carbon dioxide, mineral nutrients, and solar energy. We need differences between oceanic storage rates and leakage rates. Neither are modelled, nor constrained observationally, with anything like sufficient accuracy. However, Ref. 104 makes a strong case that the timings of the deglaciations, as distinct from their magnitudes, must be largely determined not by oceanic storage but by ice-sheet buildup.

If that's correct, the timings depend not on a small difference between ill-determined quantities — the oceanic storage rates and leakage rates — but, rather, on a single gross order of magnitude, namely the extreme slowness of ice-sheet buildup by snow accumulation. And the approximate regularity of the last four glacial cycles is then easier to understand. In any case, ocean priming seems unlikely to be slow enough to account for the full 100-millennia timespan of each cycle. But the results in Ref. 104 strongly indicate that both priming mechanisms, acting together, are crucial to explaining the huge *magnitudes* of the deglaciations.

* * *

The sensitivities considered above are, of course, the sensitivities of the real climate system to orbital changes and to carbon-dioxide injections under various circumstances, and on various timescales. They aren't the same as the sensitivities to carbon dioxide discussed in, for instance, the IPCC reports.

The reports base their sensitivities on climate-model predictions, which correspond to thought-experiments under artificial constraints. Those constraints need to be counted among the shortcomings of the models. In the models, important parts of the system including the ice sheets, and methane trapped in ice, are often held fixed in an artificial and unrealistic way. Moreover, when looking at sensitivity, attention is often confined to global-mean atmospheric temperatures — to 'global warming'. That distracts attention from the other critical aspects of climate change, including ocean heat content and the probabilities of weather extremes.

One of the critical aspects today is sea level. From a geopolitical perspective, a metre of sea level rise is large, to put it mildly. But a metre is only a tiny fraction of the more than a hundred metres by which sea levels rose between 20 millennia ago and today. It's only a tiny fraction of the further 70 metres or so by which they'd rise if all today's land-based ice sheets were to melt, taking us into a new Eocene. It's overwhelmingly improbable that an atmospheric carbon-dioxide buildup twice as large as the natural range, let alone six times as large, would fail to cause substantial sea-level rise by the end of this century. To produce a metre of global-average sea level rise, it would be enough to melt only about 5% of today's Greenland ice plus about 1% of today's Antarctic ice. After the penultimate deglaciation around 130 millennia ago, marked 5 in Figure 3, sea levels were probably *several* metres higher than today.[96]

What will actually happen to sea levels this century is one of the greatest uncertainties we now face, because of the difficulties of ice-flow modelling already emphasized. Until recently, IPCC predictions were saying that, at the end of this century, sea levels should be less than a metre higher than today. But studies of real, complex ice flows suggest that that's yet another model underprediction.[105, 106]

Another critical aspect is that our carbon-dioxide injections have *cumulative* effects on the Earth's life-support system.[96, 107] Cumulativeness means that the effects depend mainly on the total amount that's been injected. That is why climate scientists now talk about a carbon 'budget', meaning the amount remaining to be injected before the risk is considered unacceptable. By any standard of unacceptability, that 'budget' is far smaller today than the amount of carbon in fossil-fuel reserves already located,[107–110] let alone in future reserves from fossil-fuel prospecting.

The effects on the Earth's life-support system are multifarious. Besides weather extremes and rising sea levels, many other parts of the system are being impacted — not least the biodiversity that could help the system to adapt to future changes as well as helping, for instance, human medical science. Among the effects are biodiversity destruction in ocean ecosystems and food chains, such as those in coral reefs, by rising water temperatures and by the acidification that results from carbon dioxide going into the oceans. Many species have already become extinct, on land as well as in the oceans.

From a risk-management perspective it would be wise to assume that the climate-system amplifier is becoming more sensitive than it was in the pre-industrial past, and more sensitive than indicated by the climate-prediction models with their artificial constraints. The risks and uncertainties from ill-understood factors increase as the system moves further and further away from its best-known states, those of the past several hundred millennia. And increases in sensitivity can lead to tipping points.

There are at least five reasons to expect increasing sensitivity, the first of which is the complex ice-flow behaviour already referred to.

A second reason — the only one well represented in the models — is the gradual loss of sea ice in the Arctic, increasing the area of open ocean exposed to the summer sun. In northern midsummer, the 24-hour-averaged insolation increases with latitude, reaching its maximum value at the north pole, because of the Earth's 23.4° tilt. And the dark open ocean absorbs solar heat faster than the white sea ice. That's a strong positive feedback, technically called 'ice–albedo feedback', whose effects have been apparent for many years now, with the Arctic warming much faster than the rest of the planet.

A third reason is that the Arctic might also experience a tipping point involving a new and sudden loss of most of its sea ice, through a fluid-dynamical mechanism like that of the Dansgaard–Oeschger warmings. As pointed out in Ref. 71, the Arctic Ocean's uppermost layers could be destabilized by an inflow of increasingly warm subsurface Atlantic water. At first the inflow stays below the uppermost layers, which are less salty and more buoyant. But when the inflow becomes warm enough and buoyant enough, it can break through to the surface and melt the sea ice very quickly. Accelerated melting of the Greenland ice sheet, and accelerated sea-level rise, would almost certainly follow.

A fourth reason is that upper-ocean acidification could become a runaway process, through a positive feedback mechanism within the carbon cycle pointed out by Professor Daniel Rothman of MIT. Deep-ocean carbon storage via sinking plankton corpses could be drastically reduced when plankton with carbonate shells become extinct. A brief discussion is given in the first of Refs. 69. The carbonate shells cannot survive too much acidification. There's evidence that tipping points of this kind were involved in past mass extinctions, and that another might occur in the near future.

A fifth reason for increased sensitivity, again with clear tipping-point potential, is the existence of what are called methane clathrates, or frozen methane hydrates. They consist of natural-gas methane trapped in ice instead of in shale.

There are large amounts of clathrates buried on land and under the oceans, probably dwarfing conventional gas reserves although the precise amounts are uncertain.[108] As the climate system moves further beyond pre-industrial conditions, increasing amounts of clathrates will melt and release methane gas, some of which will be oxidized locally but some of which will get into the atmosphere, adding to the current leakage rates from shale-gas 'fracking' and other parts of the world's fossil-fuel infrastructure. None of those leakage rates are well quantified, but all of them seem to be increasing.[109, 110]

Methane in the atmosphere can jolt the climate system toward warmer states because for today's Earth it's an extremely powerful greenhouse gas. Its greenhouse-warming contribution per molecule is greater, by a factor of several tens, than that of the carbon dioxide to which it's subsequently converted within a decade or two (e.g. Ref. 95, Secs. 4.5.4 and 8.6.2). So the melting of clathrates, in particular, is yet another positive feedback making the climate-system amplifier more sensitive. Its magnitude is highly uncertain, but it could become strong enough to give rise to another tipping point as warming continues over the coming decades and centuries.

And there are other possible tipping points, including instability of ice flow in the Thwaites Glacier. In a worst-case scenario, with a succession of tipping points — and, to re-emphasize, there's no way to rule out such scenarios with complete confidence — it's even possible that the system might go into runaway climate change toward a hothouse state like that of the early Eocene around 56 million years ago.[111] Such a transition would take a number of centuries, probably, but again the timespan is very uncertain if only because ice-flow modelling is so uncertain. The end result would be first that there'd be no great ice sheets at all, even in Antarctica, second that sea levels would be around 70 metres higher than today, and third that the strongest thunderstorms would almost certainly be far stronger than today, and probably the cyclonic storms as well.

As already mentioned, the Clausius–Clapeyron relation says that air can hold around six or seven percent more weather fuel for each degree Celsius of temperature increase; and a thunderstorm surrounded by more weather fuel pulls the fuel in faster and reaches a greater peak intensity sooner. The geology of the early Eocene shows clear evidence of storm flood events and massive soil erosion.[112] And it was just then — surely no accident — that some land-based mammals began to migrate into the oceans. Within several million years, some of them had evolved into mammals that were fully aquatic, precursors to today's whales and dolphins. Selective pressures from extremes of surface storminess can explain those extraordinary evolutionary developments. They could well have begun with hippo-like behaviour, in which the water served only as a refuge at first.[113]

The hothouse climate of the early Eocene must have depended on high values of atmospheric carbon dioxide. Among the main non-condensing greenhouse gases that occur naturally, only carbon dioxide is chemically stable enough. For the early Eocene we have only rough estimates. Extremely high values, thousands of ppmv, are within the range of the estimates and are to be expected from large-scale volcanic activity. Past episodes of volcanic activity were sometimes far greater and more extensive than anything within human experience. Probably relevant is the episode called the North Atlantic Igneous Province, whose timing overlapped the onset of the early Eocene.

One might wonder whether the Sun's power output had a role as well. Might the early Eocene Sun have been stronger than today? With high confidence we can say that the answer is almost certainly not. The Sun was almost certainly a bit *weaker* than today, by about half a percent. That's small by comparison with the effect of carbon dioxide, but the point is worth noting.

There's a long-term upward trend in the Sun's power output, by roughly one percent per hundred million years. The trend is due to consumption of the Sun's main nuclear fuel, hydrogen, deep within its interior, along with hydrostatic adjustment between the deep interior and the visible surface, the photosphere. The solar models describing that process have become extremely secure — very tightly cross-checked — especially now that the so-called neutrino puzzle has been solved. Even before that puzzle was solved some years ago, through revised particle-physics models, state-of-the-art solar models were tightly constrained by a formidable array of observational data including precise data on the Sun's acoustic vibration frequencies, called helioseismic data. The same solar models are now known to be consistent, also, with the measured fluxes of different kinds of neutrino, subatomic particles streaming outward from the nuclear reactions that consume the Sun's hydrogen. The neutrino measurements cross-check the models in a completely independent way.

There is also some short-term variability in the Sun's behaviour. It comes mostly from variability in sunspots and other magnetic activity near the photosphere. It takes place mostly over timescales of years, decades, and centuries. Such activity is a by-product of the turbulent, boiling fluid motion caused by thermal convection in the layers beneath the photosphere, generating magnetic fields by what's called dynamo action. The variability is now known to have climatic effects distinctly smaller than the effects of carbon-dioxide injections to date, and very much smaller than those to come, let alone those in the early Eocene.

The climatic effects of solar magnetic activity include a response to the resulting variability in the Sun's power output. That variability is typically less than 0.1%, much less than half a percent. In addition, there are small and subtle effects from variability in the Sun's ultraviolet radiation, which is absorbed mainly at stratospheric and higher altitudes when the ozone layer is intact. The main points are well covered in Refs. 114 and 115. Controversially, there might be an even more subtle effect from cloud modulation by cosmic-ray shielding, which varies with solar magnetic activity. But along with their smallness, the timing of all these effects, which for the most part wax and wane every 11 years or so in what's called the 'solar cycle', implies that they have little relevance to any of the longer-term climate changes I've been discussing.

Together with the possibility of a new Eocene or hothouse Earth — one might call it the Eocene syndrome — we must also consider what might similarly be called the Venus syndrome. Not only might we get runaway climate change into an Eocene-like state, but possibly an even more drastic runaway into a state like that of the planet Venus. Venus has molten-lead surface temperatures and a nearly pure carbon-dioxide atmosphere and, of course, no oceans at all. If Venus ever had oceans, they've long since boiled away. This more drastic sense of the word 'runaway' depends on going through a stage in which the atmosphere is mostly water vapour.[95] That's still the sense most often encountered in the technical literature.

Here, however, we can be more optimistic. Even if the Earth were to go into a new Eocene, perhaps after several centuries, the Venus syndrome would be unlikely to follow, on today's best estimates. Such estimates are technically tricky — for one thing depending on poorly known cloud-radiation interactions — but they point fairly clearly toward sub-Venus-syndrome conditions even when storminess is neglected (e.g. Sec. 4.6 of Ref. 95). And the storms increase the safety margin, so to speak, by transporting heat and weather fuel away from the tropics.[116] So, whatever happens to storm-devastated, inundated human societies, life on Earth would probably carry on much as it did in the Eocene, perhaps with some more animals taking to the seas.

* * *

Coming back to our time in the twenty-first century, let's take a closer look at the storminess issue for the next few decades, focusing on the so-called temperate climates in middle latitudes. Once again, the Clausius–Clapeyron relation is basic. Global warming is global fuelling and, as well as stronger peak thunderstorms in summer, we can expect stronger peak storminess in winter, from extratropical cyclones or depressions. Weather fuel is known to be important for the intensification of such cyclones. That's why the strongest ones often form over the oceans, in what meteorologists call 'explosive' marine cyclogenesis. So, as more weather fuel becomes available, the strongest extratropical cyclones can be expected to become increasingly strong, in terms of damaging rainfall and damaging winds.

I must emphasize that I'm talking about the strongest and most damaging cyclones, and not about the average cyclone — just as I talked earlier about the strongest and most damaging thunderstorms and not the average thunderstorm. Both are outside the scope of the climate prediction models because the most extreme behaviours involve small spatial scales, which the models can't resolve. For extratropical cyclones a counter-argument is sometimes made that, because the cyclones are embedded in changing surroundings, as the climate changes, the weather-fuel effect need not predominate. There's a technical term, 'baroclinicity', used to describe horizontal temperature gradients in the surroundings; and climate change may lead to reduced baroclinicity which, other things being equal, would itself lead to reduced cyclogenesis. That, however, is an argument more applicable to large-scale average conditions than to local extreme conditions.

Local extremes in baroclinicity aren't the only issues that would need to be considered in a full assessment. Extratropical cyclones and jet streams feed off each other in a complex fluid-dynamical interplay, strongly dependent on the Earth's rotation. It's a subtle interplay that for one thing stops jet streams from spreading like ordinary, candle-blowing-out air jets. Instead, it keeps them narrow as they meander over thousands of kilometres. And that in turn is part of a complex cause-and-effect jigsaw extending out to global scales. In the technical literature the long-range causal influences are called 'teleconnections'. They too depend on the Earth's rotation and have an interplay with, for instance, jet stream meandering on timescales of days to weeks, and with phenomena such as El Niño on longer timescales. El Niño involves coupling between the atmosphere and the upper ocean, especially the tropical Pacific, leading to large-scale fluctuations in tropical sea-surface temperatures and in storm statistics on multi-year timescales. Vast amounts of heat are transferred back and forth between atmosphere and ocean, on those timescales, creating what misleadingly look like global-warming 'hiatuses' when attention is confined to atmospheric temperatures. Such a 'hiatus' occurred in the early 2000s when it was used by the climate disinformers to declare that global warming had ended. Many aspects of the jigsaw are still not accurately represented in the climate prediction models, despite recent improvements.

With a full assessment still out of reach it seems to me reasonable to say, for now, that the availability of more weather fuel is likely to be making the whole climate system more active, with larger fluctuations, not only in thunderstorm extremes but in many other ways besides. As mentioned earlier, larger fluctuations generally means larger in both directions. Examples include severe droughts in some places, and severe flooding in others due not to thunderstorms but to persistent heavy rain fed by moisture-laden low-altitude 'conveyor belts', or 'atmospheric rivers'.

Further examples include the recent large-scale weather fluctuations from amplified meandering of high-altitude jet streams, such as the enormous meander that brought unusual snowfall and severe cold to Europe, starting in late February 2018. In news reports it was named 'The Beast from the East'. Another such meander brought a severe cold air outbreak over the USA in January–February 2019. Again, a full assessment is out of reach, but such cold events are worth noting, it seems to me, if only because of the way the disinformers have used them to reiterate that climate science is wrong, and that global warming has ended. In February 2021, there was a cold outbreak over the USA that was similar but still more severe, with crippling snowfall and burst water pipes as far south as Texas. December 2022 saw yet another such outbreak that was even worse, causing dangerous cyclonic blizzards, whiteouts, and dozens of fatalities over large areas of North America. In June–July 2021, an extreme heat wave with firestorms engulfed western parts of North America, thanks to another huge jet-stream meander, this time in the opposite direction, poleward rather than equatorward. It seems clear that these fluctuations are indeed getting larger — larger in both directions.

Although I've emphasized the shortcomings of the climate prediction models, I should again acknowledge that they're an important part of our hypothesis-testing toolkit, when used within their limitations. Up-to-date summaries of today's state of the art, and future prospects, in climate modelling can be found in Refs. 117 and 118. But despite the many improvements in the models, the challenge of bringing them closer to the problem of weather extremes remains one of the toughest of all the challenges facing climate science.

Today's *operational weather-forecasting models* are getting better at simulating some of the extremes. That's mainly because of their far finer spatial resolution, obtained at great computational cost per day's simulation.

The computational cost means that such models can't be used for direct simulations of climate change over centuries. However, in recent years there have been studies making indirect use of local, fine-resolution operational forecasting models. Such models now have enough spatial resolution to begin to describe individual thunderstorms, and can be embedded in more coarsely resolved simulations of the surrounding climate. One such study was that of Elizabeth Kendon and co-workers at the UK Meteorological Office.[119] The local model covered the southern UK and was embedded in a coarser climate prediction for the years 2090–2110. That allowed the local model to be run long enough to produce significant statistics. The results showed that with more weather fuel available both the magnitudes and the frequencies of extreme summertime thunderstorms increased. The extremes were greater than would be predicted from the Clausius–Clapeyron relation alone, but in line with expectations from the robust positive feedback already mentioned — more weather fuel being pulled in faster to give storms a quicker and greater peak intensity.

As computer power increases there will be many more such studies, at increasingly fine resolutions, beginning to describe more accurately the statistics of extreme rainstorms, snowstorms, and windstorms. Extreme winter cyclones in middle latitudes are spatially more extensive than individual thunderstorms, and are better simulated, but again only by the operational forecasting models and not by the climate prediction models. Tropical cyclones like Idai and Dorian pose their own peculiar modelling difficulties but, as with other kinds of weather extremes, the evidence suggests that the most destructive tropical cyclones have tended to be underpredicted rather than overpredicted.[120]

* * *

Dear reader, before taking my leave I should be careful to acknowledge the imperfections of the amplifier metaphor. We do need a way of talking about the climate system that recognizes some parts of it as being more sensitive than others. However, in portraying the system as an amplifier we also need to recognize its many intricately-linked components operating over a huge range of timescales, some of them out to multi-decadal, multi-century, multi-millennial, and even longer, as we've seen. And the climate-system amplifier would be pretty terrible as an audio amplifier not only because of its highly variable input sensitivities — in technical language, strong 'nonlinearity' — but also because it has so much chaotic internal noise and self-generated variability, on so many timescales. Some of the system's chaotic internal fluctuations have the character of tipping points, for instance the Dansgaard–Oeschger events described in Refs. 69 and 71 — warning us, as already said, that a similar tipping point might soon cause the sudden disappearance of today's Arctic sea ice, followed almost certainly by an acceleration in the melting of the Greenland ice sheet.

All these complexities have helped the climate disinformers because, from the many signals within the system, showing its chaotic variability in time, one can always cherry-pick some that seem to support practically any view one wants — especially if one replaces insights into the workings of the system, as seen from several viewpoints, by simplistic, post-hoc-fallacy arguments that conflate timing with cause and effect, as with the cock's crow causing the sunrise. Chaotic variability provides many opportunities to cherry-pick data segments, showing what looks like one or another trend or phase lag and adding to the confusion about causal relations on different timescales.

If you see two signals within a complex system, with one of them looking like a delayed version of the other, then nothing can be deduced about causality from that information alone. Since there are many links within the complex web of cause and effect, it's possible that the two signals you're looking at are separately influenced by something else in the system.

The point about cherry-picking is so important that I'd like to illustrate it by returning briefly to what happened during the last deglaciation. A compilation by my colleague Dr Luke Skinner, who is an expert on the observational evidence, gives us a glimpse of the actual complexity we're dealing with. He presents graphs showing *six* of the signals, reproduced here as Figure 20, in which time runs from right to left as before:

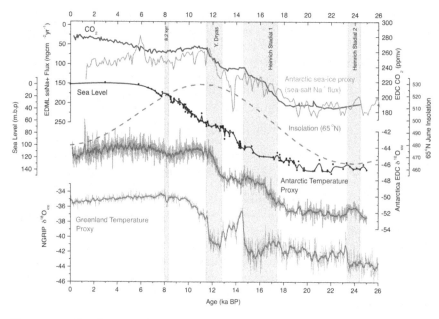

Figure 20: Graphs recording some of the major events of the last deglaciation, courtesy of Dr Luke Skinner. Time, in millennia, runs from right to left up to the present day. 'EDML' and 'EDC' refer to data from ice cores taken at high altitude on the East Antarctic ice sheet, respectively in Dronning Maud Land and at a location called Dome C; 'NGRIP' refers to an ice core from a high-altitude location in Greenland. The smooth dashed graph shows the midsummer insolation at 65°N, reproducing the leftmost peak in Figure 19. Atmospheric carbon dioxide (CO_2) from EDC is shown as the dark graph at the top. The two graphs at the bottom show oxygen-isotope ratios believed to be rough indicators, or proxies, for air temperature at the time when snow fell on to the ice; the corresponding temperature ranges are of the order of several degrees Celsius. The light graph near the top marked 'Antarctic sea-ice proxy' is a measured quantity believed to be a proxy for the variability of sea ice extent around the Antarctic continent,[121] with more sea ice when the graph dips down, as on the right. Further information on events during the deglaciation can be found, for instance, in Refs. 98–104.

In particular, the dark graph at the top shows atmospheric carbon dioxide from an Antarctic ice core, while the graph at the bottom shows isotope data giving an indication of air temperatures in Greenland. The typical disinformers' ploy — cherry-picking data segments — would in this case start by restricting attention to times between 21 and 17½ millennia ago. Then attention would be restricted to the carbon dioxide and Greenland temperature graphs. The disinformer would then say "Look, temperatures were rising well before carbon dioxide started rising; so temperature rise causes carbon dioxide rise rather than *vice versa*." Notice the dichotomization, shutting off further thinking.

A better commentary on the six graphs — informed by our best, though still very imperfect, understanding of how the climate system works, including ice-flow dynamics[98–106] — would be as follows. The graphs point to a complex sequence of events in which the initial increase in midsummer insolation at 65°N (smooth dashed graph, cf. leftmost peak in Figure 19) increased the flow of northern land-based ice and meltwater into the sea. This began a long-drawn-out rise in sea level (graph with dots). Thanks to the ever-increasing insolation, reinforced by a multi-millennial delay in the 'isostatic rebound' of the Earth's crust[104] — keeping the top surface of the ice at lower, warmer altitudes — summertime melting rates increased further over successive millennia. Under the influence of meltwater buoyancy and the Earth's rotation, the flow of meltwater into the sea induced changes in the Atlantic overturning and Southern Ocean circulations, and a reduction in Antarctic sea-ice extent (light graph near top). Together those changes led to the release of carbon dioxide from a well primed deep-ocean storage reservoir, entering the atmosphere mainly through expanded ice-free areas of the Southern Ocean. The release of carbon dioxide acted as a positive feedback. Its greenhouse effect added substantially to the warming and ice-melting effects, boosting sea-level rise still further.[98, 104]

I can say more, too, about why we can trust the ice-core records of atmospheric carbon dioxide, such as the top graph in Figure 20 and the bottom graph in Figure 3. The ice-core records count as extremely hard evidence thanks to the meticulous cross-checking that's been done — for instance, by comparing results from different methods of extracting the carbon dioxide trapped in ice, by comparing results between different ice cores having different snow accumulation rates, and by comparing ice-core data with the direct atmospheric measurements that have been available since 1958. We really do know with practical certainty the past as well as the present atmospheric carbon-dioxide concentrations, with accuracies of the order of a few percent, as far back as about 800 millennia even though not nearly as far back as the ice-free Eocene. Because of the chemical stability of carbon dioxide as a gas — in unmelted ice as well as in the atmosphere — it's well preserved in Antarctic ice cores, and well mixed throughout most of the atmosphere. Its atmospheric concentration in ppmv hardly varies from north to south. So the Antarctic ice-core values can be read as accurate global values.[122]

I should also explain why we can have high confidence in what I've said about sea levels, that they went up and down by amounts of the order of a hundred metres or more, at times during the glacial cycles. We know about past sea levels from several hard lines of geological evidence, including direct evidence from old shoreline markings and coral deposits. It's difficult to allow accurately for the isostatic deformation of the Earth's crust by changes in ice and ocean mass loading, and for tectonic effects generally. But the errors from such effects are likely to be of the order of metres only. Such errors hardly affect the overall picture. And an independent cross-check comes from oxygen isotope records, reflecting the fractionation between light and heavy oxygen isotopes when water evaporates from the oceans and is deposited as snow on the great ice sheets. That cross-check is consistent with the geological shoreline estimates.[123]

* * *

So we've been driving in the fog, scientifically speaking, but the fog is now clearing. We're reaching a better understanding of the risks — including the risks from weather extremes, from sea-level rise, and from the massive uncertainty over tipping points.[107] And the politics is now, at long last, beginning to change significantly. The infocalypse[21] notwithstanding, the disinformation campaigns are no longer the *overwhelming* political influence that they were as recently as a decade ago. They seem now, at long last, to be meeting the same fate for climate as they did for the ozone hole, and for tobacco and lung cancer.[24] All three cases show the same pattern: disinformation winning at first, then defeated by a strengthening of the science along with a wave of public concern powered by real events.[124]

Of course climate, as such, isn't the only challenge ahead. There's the destruction of biodiversity, and the long-studied evolution of zoonotic viruses such as those giving rise to HIV–AIDS, Ebola, and the SARS–MERS–COVID diseases. The recent COVID pandemic, or something like it, had long been anticipated by the scientists concerned. The pandemic has reminded us that playing havoc with the Earth's life-support system, invading its ecosystems and their virus populations, might look lucrative but can actually be costly in human terms — and likely more so in future.

And there's the evolution of antibiotic resistance in bacteria. There's the threat of asteroid strikes. There's the enormous potential for good or ill in new nanostructures and materials,[55] in artificial intelligence, and in gene editing.[125] There are the threats from cybercrime, from cyberwarfare, and from automated kinetic warfare — 'Petrov's nightmare', one might call it. All these things demand clear thinking, good science and engineering, and good risk management. The number of ways for things to go wrong is combinatorially large, sometimes with huge unintended consequences including the assaults on democracy, on science, and on mental health via the social media, with their lightning-speed artificial intelligences that aren't yet very intelligent.[21] So I come back to my hope that good science, which in practice means open science, with its powerful ideal and ethic, its humility and its respect for evidence, and its ability to cope with the unexpected, will continue to survive and prosper despite all the forces ranged against it.

After all, there are plenty of daring and inspirational examples. One of them is the continuing work of the open-source software community,[17] and another was Peter Piot's work on the HIV–AIDS pandemic, not only on the science itself but also on persuading pharmaceutical corporations to make antiviral drugs affordable worldwide. And there was the response to the COVID-19 pandemic, with the advent of revolutionary new MRNA-based[125] vaccines against the virus. Yet another example was the human-genome story, reaching its climax around the turn of the century. There, the scientific ideal and ethic kept the genomic information available to open science including medical and virological research, in the teeth of powerful efforts to lock it up and monopolize it commercially.[26]

When one contemplates not only human weakness but also the vast resources devoted to individual profit by fair means or foul — and to disinformation — one can't fail to be impressed that good science ever gets anywhere at all. That it has done so again and again is to me, at least, very remarkable, and inspirational. We humans can be much, much smarter and wiser than simplistic evolutionary theory allows.

The ozone-hole story, in which I myself was involved professionally, is another example. The disinformers tried to discredit everything we did, using the power of their well-resourced political weapons. What we did was seen — as with tobacco and lung cancer — solely as a threat to share prices and profits. And yet the science, including all the cross-checks between different lines of evidence both observational and theoretical, became strong enough, adding up to enough in-depth understanding, despite the complexity of the problem, to defeat the disinformers in the end. So we now have the Montreal Protocol to limit ozone-depleting chemicals, in a new symbiosis between market forces and regulation. That too was inspirational. Surely Adam Smith would have approved.[2, 25, 40] And it has bought us a bit more time to deal with climate, because the ozone-depleting chemicals are also potent greenhouse gases. If left unregulated, they would have accelerated climate change still further.

And on climate itself, as already said, we seem at long, long last to be close to what looks like a similar political turning point. The risk-management side of the problem was clearly highlighted some time ago.[126] The Paris climate agreement of December 2015, and now the schoolchildren's and other mass movements, allow us to hope for another symbiosis between market forces and regulation, especially now that low-carbon 'renewable' energy has become so much cheaper than fossil-fuel energy.[124] That's a powerful push toward a 'green' economy and many new jobs.[124, 127]

My colleague Myles Allen, who was instrumental in pointing out the cumulativeness of carbon-dioxide injections and its consequence, the carbon 'budget' — the need to limit *total* future injections, or emissions[107] — has suggested that a new symbiosis could emerge from recognizing the parallel between the fossil-fuel power industry and the nuclear power industry. For a long time now, it has been unthinkable to plan a nuclear power plant without including the waste-disposal engineering and its costs. Progress, then, could come from recognizing the same thing for the fossil-fuel industry. Then carbon capture and storage,[128] allowing safe fossil-fuel burning as well as, for instance, safe methane-to-hydrogen conversion and thence safe aircraft fuelling, could become a reality at scale by drawing on the fossil-fuel industry's engineering skills.

And the limitless burning of fossil fuels without carbon capture and storage is increasingly seen as risky even in purely financial terms. It's seen as heading toward stranded assets and what's been called the bursting of the shareholders' carbon bubble. James Thornton's ClientEarth has been bringing successful lawsuits to stop high-carbon and other high-pollution investments in many countries, arguing *irresponsibility to shareholders* as well as to society. And more and more ideas are coming to maturity about diversifying and driving down the cost of low-carbon energy. Beyond renewables[124] an interesting example is the Rolls Royce consortium on 'small modular reactors' for nuclear energy, produced cost effectively by state-of-the-art manufacturing techniques that avoid the spiralling costs of the large, one-off reactor projects.[129] And now, at long last, we're seeing the electrification of personal transport, a beautiful and elegant technology and another step in the right direction.

As regards good science in general an important factor in the human-genome story, as well as in the ozone-hole story, was a policy of open access to experimental and observational data. That policy was one of the keys to success. The climate-science community was not always so clear on that point, giving the disinformers further opportunities. However, the lesson now seems to have been learnt.

I don't think, by the way, that everyone doubting climate science is dishonest. Honest scepticism is crucial to science; and I wouldn't question the sincerity of colleagues I know personally who feel, or used to feel, that the climate-science community got things wrong. Indeed, I'd be the last to suggest that that community, or any other scientific community, has never got anything wrong, even though my own sceptical judgement is that, as already argued, the climate-science consensus in the IPCC reports is mostly right apart from *underestimating* the problems ahead, including sea-level rise — partly because the climate prediction models have underestimated them, and partly for fear of being accused of alarmism and scaremongering.

It has to be remembered that unconscious assumptions and mindsets are always involved, in everything we do and think about. The anosognosic patient is perfectly sincere in saying that a paralyzed left arm isn't paralyzed. There's no dishonesty. It's just an unconscious thing, an extreme form of mindset. Of course, the *professional* art of disinformation includes the deliberate use of what sales and public-relations people call 'positioning' — the skilful manipulation of other people's unconscious assumptions, related to what cognitive scientists call 'framing'.[41]

As used by professional disinformers the framing technique exploits, for instance, the dichotomization instinct to evoke the unconscious assumption that there are only two sides to an argument. The disinformers then insist that their 'side' merits equal weight, as with flat Earth versus round Earth. That's sometimes called 'false balance'. In this and in many other ways the disinformers spread confusion, exploiting their deep knowledge of the way perception works. It's inspirational, therefore, to see the *climate* disinformers, despite their mastery of such techniques, now facing defeat.

It often takes a younger generation to achieve Max Born's 'loosening of thinking', exposing unconscious assumptions and making progress. Science, for instance, has always progressed in fits and starts, always against the odds, and always involving human weakness alongside a collective struggle with erroneous mindsets exposed, usually, through the efforts of a younger generation. The great population geneticist J. B. S. Haldane famously caricatured it in four stages: (1) This is worthless nonsense; (2) This is an interesting, but perverse, point of view; (3) This is true, but quite unimportant; (4) I always said so. The push beyond simplistic evolutionary theory is a case in point. So is the acceptance of Bayesian statistics.[83]

So here's my farewell message to young scientists, and to any young person who cares about a civilized future. You have the gifts of intense curiosity and open-mindedness. You're willing and able to think logically. You have the best chance of spotting inconsistency and misinformation, and of detecting unconscious mindsets and dichotomizations. You have enormous computing power at your disposal. You have brilliant programming tools, and observational and experimental data far beyond my own youthful dreams of long ago. You have a powerful new mathematical tool, the Bayesian probabilistic 'do' operator, for distinguishing correlation from causality in complex systems.[22] You'll have seen how new insights from systems biology — transcending simplistic evolutionary theory — have opened astonishing new pathways to technological innovation[18, 55, 125] as well as deeper insights into our own human nature, as I've tried to sketch in Chapter 3. And above all you know how to disagree without hating — how to view things from more than one angle and how to argue over the evidence, not to score personal or political points, but to enjoy exploring things and to reach toward an improved understanding.

Whatever your fields of expertise, you know that it's fun to be curious, to be open to the unexpected, and to be surprised when you discover how something works. It's fun to do thought-experiments and computer experiments. It's fun to develop and test your in-depth understanding, the illumination that can come from looking at a problem from more than one angle. You know that it's worth trying to convey that understanding to a wide audience, if you get the chance. You know that in dealing with complexity you'll need to hone your communication skills in any case, if only to develop cross-disciplinary collaboration, the usual first stage of which is jargon-busting — as far as possible converting turgid technical in-talk into plain, lucid speaking.

So hang in there. Your collective brainpower will be needed as never before.

References and endnotes

1. Many of Darwin's examples of cooperative behaviour, within and even across different social species, can be found in Chapters II and III of his famous 1871 book *The Descent of Man*. Further such examples were noted by another great 19th-century thinker, Peter Kropotkin. Darwin's greatness as a scientist shows clearly in the meticulous attention he pays to observations of actual animal behaviour. The animals observed include primates, birds, dogs, and ruminants such as cattle. See especially pages 74–84 in the searchable full text of his book, which is available online at http://darwin-online.org.uk/content/frameset?pageseq=1&itemID=F937.1&viewtype=text. The contrary idea that competition between individuals is all that matters — often wrongly attributed to Darwin — goes back further. It shows up in the writings of, for instance, the philosopher Thomas Hobbes. In his 1651 book *Leviathan*, Hobbes famously declared that human nature is innately vicious, and that only authoritarian dictatorship can save us from a dog-eat-dog life that's "nasty, brutish and short", or "red in tooth and claw" as Alfred, Lord Tennyson later put it. Despite his great learning, Hobbes could not have known much about actual hunter-gatherer societies, whose individuals show many instinctively compassionate and cooperative modes of behaviour that Hobbes, it seems, would have found surprising.[2, 10, 11]

2. See for instance Collier, P. and Kay, J., 2020: *Greed is Dead: Politics After Individualism*, Penguin, Allen Lane. John Kay and Paul Collier are British experts on business and economics. Their book discusses the damage

done to human societies by the idea that competition between individuals, 'winner versus loser', is all that matters. See also economist Noreena Hertz's 2020 book *The Lonely Century: Coming Together in a World That's Pulling Apart*, Sceptre. The damage has been done *not only* via the gross economic inequality and the resulting 'anti-trickle-down' impoverishment of the poorest, from what economists now call neoliberalism, or market fundamentalism[25] — on the so-called political right alongside 'possessive individualism' or 'greed is good' — *but also*, in various ways, across a wide range of persuasions on the so-called political left and sadly, in addition, by strengthening the subcultures that tolerate or advocate violence toward individuals, including sexual violence. However, as noted by Kay, Collier, and Hertz, Darwinian evolutionary theory does not, in fact, say that competition between individuals is all that matters. On the contrary, evolution has given us not only the violent and greedy sides of human nature but also the friendly, cooperative, build-it-together, *compassionate* sides[10, 11] — 'ubuntu' in the sense championed by Desmond Tutu, the instinctive feeling that 'I am because we are' — the denial of which, and the resulting loneliness, has been devastating for physical as well as for mental and social health. That conclusion is supported by today's cutting-edge biology, as I'll show in my Chapter 3. Kay, Collier, and Hertz are also very clear on Adam Smith's views which, contrary to most opinions today, recognized as important that there's far more to human nature than simple greed or selfishness.[40] The point has been underlined by the young people who in 2020 volunteered for COVID-19 'challenge trials' — deliberate infection under controlled conditions — taking personal risks to help us all by accelerating scientific understanding, and vaccine development, in the arms race against an invisible, agile, and rapidly mutating enemy.[125]

3. Monod, J., 1971: *Chance and Necessity*, Glasgow, William Collins, beautifully translated from the French by Austryn Wainhouse. This classic by the great molecular biologist Jacques Monod — one of the sharpest and clearest thinkers science has ever seen — highlights the "more than two million years of directed and sustained selective pressure" (Chapter 7, page 124) arising from the co-evolution of our ancestors' genomes and cultures, entailing the gradual emergence of proto-language and then language because (page 126) "once having made its appearance, language, however primitive, could not fail... to create a formidable and oriented selective

pressure in favour of the development of the brain" in new ways — to the advantage of "the groups best able to use it". The possibility of such group-level selective pressure is still controversial and I'll return to it in my Chapter 3, where in addition another topic from Monod's book will come up, namely the regulatory 'biomolecular circuits' overlying the genome. Those circuits depend on transistor-like components called 'allosteric enzymes'. Monod won a Nobel prize for his pioneering work on such enzymes.

4. Tobias, P. V., 1971: *The Brain in Hominid Evolution*, New York, Columbia University Press. Phillip Tobias's famous research on palaeoanthropology led him to recognize the likely interplay of genomic and cultural evolution, despite the vast disparity of timescales. He writes that "the brain–culture relationship was not confined to one special moment in time. Long-continuing increase in size and complexity of the brain was paralleled for probably *a couple of millions of years* [my emphasis] by long-continuing elaboration... of the culture. The feedback relationship between the two sets of events is as indubitable as it was prolonged in time." Evidence in further support of this picture will be summarized in my Chapter 3.

5. I began with examples of unconscious assumptions I've encountered in my own scientific research. Their unconscious nature was very clear because making them conscious exposed them as self-evidently wrong, indeed silly. They're discussed in a paper of mine, "On multi-level thinking and scientific understanding," *Advances in Atmospheric Science* **34**, 1150–1158 (2017).

6. Kahneman, D., 2011: *Thinking, Fast and Slow*, London, Penguin. Daniel Kahneman's book, based on his famous research with Amos Tversky on psychology and economics, provides deep insight into many unconscious processes of great social importance. See also Ref. 41. Early in his Introduction, Kahneman gives an example of a lifesaving decision taken ahead of conscious thought. In Kahneman's words, the commander of a team fighting a domestic fire "heard himself shout, 'Let's get out of here!' without realizing why." The commander had unconsciously sensed that the floor on which they were standing was about to collapse into a bigger fire underneath.

7. Bateson, G., 1972: *Steps to an Ecology of Mind: Collected Essays on Anthropology, Psychiatry, Evolution and Epistemology*, republished 2000 by

the University of Chicago Press. The quoted sentence is on page 143, in the section on 'Quantitative Limits of Consciousness'.

8. The unconscious learning that's needed to acquire normal vision has been extensively studied. For instance, there's been a long line of research on whether anything like normal vision is acquired by individuals blinded in infancy by congenital cataracts, or opaque corneas, that are surgically removed later in life. Common to all such cases is the complete absence of normal vision immediately after surgery. The patient typically sees only a confused mess of shapes and colours. Many months after surgery, abilities much closer to normal vision can be acquired, especially by the youngest patients. Some older patients fail to acquire anything remotely like normal vision, as in classic cases of middle-aged patients, including a man called Virgil, described in Oliver Sacks's 1995 book *An Anthropologist on Mars*, New York, Alfred Knopf. In recent years, Project Prakash in India has encountered many cases of younger patients who do better. See for instance Sinha, P., 2013: "Once blind and now they see," *Scientific American* **309**(1), 48–55, and Gandhi, T., Singh, A.K., Swami, P., Ganesh, S., Sinha, P., 2017: "The emergence of categorical face perception after extended early-onset blindness," *Proceedings of the National Academy of Sciences* **114**, 6139–6143. This last study demonstrates not the ability to distinguish the faces of different individuals from one another, but only the ability to distinguish images of human faces from images of other things. That's typically acquired by children within a year or so after surgery. Oliver Sacks's book also describes the case of the painter Franco Magnani, to be mentioned in my Chapter 4.

9. See for instance Chapter 4 of Noble, D., 2006: *The Music of Life: Biology Beyond Genes*, Oxford University Press. This short and lucid book by an eminent biologist, Denis Noble FRS, clearly brings out the complexity, versatility, and multi-level aspects of biological systems, and the need to avoid extreme reductionism and single-viewpoint, hypercredulous thinking — such as saying that the genome 'causes' everything. A helpful metaphor for the genome, it's argued, is a digital music recording. Yes, reading the digital data 'causes' a musical and possibly an emotional experience but, if that's all you say, you miss the other things on which the experience depends so strongly, coming from past experience as well as

present circumstance. Reading the data into a playback device mediates or *enables* the listener's experience, rather than solely *causing* it.

10. The social commentator David Brooks has put forward a wealth of evidence that human compassion is just as natural and instinctive as human nastiness. Further evidence is offered in, for instance, Refs. 2 and 11. Brooks regards compassion as something mysterious and beyond science, because he takes for granted what I'll call 'simplistic evolutionary theory' — the dog-eat-dog, purely competitive view of biological evolution that still permeates popular culture. By contrast, a more complete view of evolution beginning with Charles Darwin's observations,[1] and summarized in my Chapter 3, easily explains what Brooks describes as "the fierceness and fullness of love, as we all experience it". Brooks describes many hundreds of people whom he's met — he calls them "social weavers" — who live for others with little material reward. As Brooks observes, they are rewarded instead by something deep in their inner nature, giving them joy in caring for others and in the spontaneous upwelling of emotions such as love and compassion. They're at peace with themselves because this inner life is more important to them than competition for power, money, and status. They radiate joy. They "seem to glow with an inner light... they have a serenity about them, a settled resolve. They are interested in you, make you feel cherished and known, and take delight in your good." All this comes naturally and without conscious effort. It does not come from being preached at moralistically. It does not come from cynical, consciously calculated 'virtue signalling'. See Brooks, D., 2019: *The Second Mountain: The Quest for a Moral Life*, Random House, Penguin.

11. Bregman, R., 2020: *Humankind: A Hopeful History*, Bloomsbury. This book by Rutger Bregman assembles many examples of human behaviour that's instinctively friendly, kind, and compassionate, outside kin or family. Bregman acknowledges Hobbesian human nastiness[1] but makes the point that such nastiness isn't all there is. He gives many examples to add to those in Ref. 10 and argues that our friendly, compassionate side must have been a key part of our ancestors' extraordinary versatility and adaptability — contrary to the dog-eat-dog view of biological evolution that still permeates popular culture, which I call 'simplistic evolutionary theory'. And he gives us a reassessment of what Hannah Arendt called "the banality of evil", with

a forensic re-examination of the case of Adolf Eichmann, as well as of the notorious psychological experiments of Stanley Milgram and others, in the light of newly unearthed evidence.

12. See for instance Pagel, M., 2012: *Wired for Culture*, Allen Lane, Penguin, Norton. In this book, a well-known evolutionary biologist, Mark Pagel FRS, describes many interesting aspects of recent cultural evolution. On human languages, though, he considers that they are all descended from a single 'mother tongue'. That's seen most clearly on page 299, within a section headed "words, languages, adaptation, and social identity". The author suggests that language and the mother tongue were invented, as a single cultural event, at some time after "our invention of culture" around 160 to 200 millennia ago (page 2). This hypothetical picture is diametrically opposed to that of Monod,[3] who along with Tobias[4] insists on a strong feedback between genomic evolution and cultural evolution over a far longer timespan, under strong selective pressures at group level (Monod,[3] pp. 126–7). Pagel's arguments assume that group-level selection is never important. I myself find Monod's arguments more persuasive, for reasons set out in my Chapter 3. Powerful cross-checks come from recent events in Nicaragua alongside laboratory experiments that directly demonstrate group-level selection.[43]

13. A typical range of views and controversies over the origins of language can be found in the collection of short essays by Trask, L., Tobias, P. V., Wynn, T., Davidson, I., Noble, W., and Mellars, P., 1998: "The origins of speech," *Cambridge Archaeological Journal* **8**, 69–94. There are contributions from linguists, palaeoanthropologists, and archaeologists. Some of the views are consistent with Monod's,[3] while others (see especially the essay by Davidson and Noble) are more consistent with the idea espoused by Pagel[12] that language was a very recent, purely cultural invention. That idea still pops up in discussions today. See also the thoughtful discussion in Ref. 61.

14. van der Post, L., 1972: *A Story Like the Wind*, London, Penguin. Laurens van der Post celebrates the influence he felt from his childhood contact with some of Africa's "immense wealth of unwritten literature", including the magical stories of the San or Kalahari-Desert Bushmen, stories that come "like the wind... from a far-off place." See also van der Post's 1961 book *The Heart of the Hunter*, Penguin, page 28, on how a Bushman told what had happened to his small group: "They came from a plain... as they put it in

their tongue, 'far, far, far away'.. It was lovely how the 'far' came out of their mouths. At each 'far' a musician's instinct made the voices themselves more elongated with distance, the pitch higher with remoteness, until the last 'far' of the series vanished on a needle-point of sound into the silence beyond the reach of the human scale. They left... because the rains just would not come..."

15. Pinker, S., 2018: *Enlightenment Now: The Case for Science, Reason, Humanism and Progress*, Penguin, Random House. In a thoughtful chapter on democracy the author discusses why, in a long-term view — and contrary to what headlines often suggest — democracy has survived against the odds and spread around parts of the globe in three great waves over the past two centuries. It seems that this is not so much because of what happens in elections (in which the average voter isn't especially engaged or well informed) but, as pointed out by political scientist John Mueller, more because it gives ordinary people the freedom to complain publicly at any time, and to be listened to by governments without being imprisoned, tortured, or killed. In other words, it's to some degree an 'open society' in the sense of Karl Popper — allowing bottom-up influence or 'accountability' — which for one thing improves the society's versatility and adaptability. (Of course, I'm assuming that 'democracy' means more than popular voting. As discussed by political scientists such as Juan Linz, in order to be more than just a sham it must include reasonably strong counter-autocratic institutions, including the separation of executive, judicial, and vote-counting powers — and above all a culture that allows disagreement without hating, hence acceptance of electoral wins by opponents.)

16. For this new cross-check (verifying a prediction made over a century ago by Albert Einstein from his general-relativity equations) see Abbott, B. P., *et al.*, 2016: "Observation of gravitational waves from a binary black hole merger," *Physical Review Letters* **116**, 061102. This was an enormous team effort at the cutting edge of high technology, decades in the making, to cope with the minuscule amplitude of Einstein's lightspeed gravitational ripples. The '*et al.*' stands for the names of over a thousand other team members. The study used three LIGO detectors (Laser Interferometer Gravitational-Wave Observatory), two in the USA and one in Italy, called Virgo. They are giant interferometers, with arms kilometres long, accommodating laser beams that can detect length changes of the order of a billion-billionth

of a metre or less. The first event was detected on 14 September 2015. A second event was detected on 26 December 2015. It was reported in Abbott, B. P., *et al.*, 2016: "GW151226: Observation of gravitational waves from a 22-solar-mass binary black hole coalescence", *Physical Review Letters* **116**, 241103. In this second event, the model-fitting indicated that one of the merging black holes probably had nonzero spin. Subsequent events have corresponded to the merging not only of black holes but also of neutron stars.

17. Valloppillil, V., 1998: *The Halloween Documents: Halloween I*, with commentary by Eric S. Raymond. This leaked document from the Microsoft Corporation, available online, records Microsoft's secret recognition that software far more reliable than its own was being produced by the open-source community, a major example being the operating system called Linux. *Halloween I* states, for instance, that the open-source community's ability "to collect and harness the collective IQ of thousands of individuals across the Internet is simply amazing." Linux, it goes on to say, is an operating system in which "robustness is present at every level" making it "great, long term, for overall stability". I well remember the non-robustness and instability, and user-unfriendliness, of Microsoft's own secret-source software during its near-monopoly in the 1990s.

18. Recent advances in understanding genetic codes include insights into how they influence, and are influenced by, the lowermost layers of complexity in the molecular-biological systems we call living organisms. Some of these advances, building on those in Refs. 3 and 9, are beautifully described in the book by Andreas Wagner, 2014: *Arrival of the Fittest: Solving Evolution's Greatest Puzzle*, London, Oneworld. A detailed view emerges of the interplay between genes and functionality, such as the synthesis of chemicals needed by an organism, and the switching-on of large sets of genes to make the large sets of enzymes and other protein molecules required by a particular functionality. And these systems-biological insights throw into sharp relief the naïvety, and absurdity, of thinking that genetic codes are blueprints governing everything, and that there is a single gene 'for' a single trait.[9] Also thrown into sharp relief is the role of natural selection *and*, importantly, of other mechanisms such as neutral mutation, in the evolution of functionality. The importance of the other mechanisms vindicates one of Charles Darwin's key insights, namely that natural selection is only

part of how evolution works. In Darwin's own words from page 421 of the sixth, 1872 edition of his *Origin of Species*, "As my conclusions have lately been much misrepresented, and it has been stated that I attribute the modification of species exclusively to natural selection, I may be permitted to remark that in the first edition of this work, and subsequently, I placed in a most conspicuous position — namely, at the close of the Introduction — the following words: 'I am convinced that natural selection has been the main but not the exclusive means of modification.'" And the research by Wagner and co-workers shows clearly and in great detail the role of, for instance, vast numbers of neutral mutations. These are mutations that have no immediate adaptive advantage and *are therefore invisible to natural selection*, yet have crucial long-term importance through building genetic diversity.[44] The research gives insights into the extraordinary robustness of biological functionality, insights that have practical technological implications. For instance, they've led to new ways of designing, or rather discovering, robust electronic circuits and computer codes. Further such insights come from recent studies of self-organizing or self-assembling structures, as for instance with the 'murmurations' of flocks of starlings, and the structures emerging in crowds of 'swarm-bots'.

19. Vaughan, M. (ed.), 2006: *Summerhill and A. S. Neill*, with contributions by Mark Vaughan, Tim Brighouse, A. S. Neill, Zoë Neill Readhead, and Ian Stronach, Maidenhead, New York, Open University Press/McGraw-Hill. Summerhill School is a boarding school founded in 1921 by Alexander Sutherland Neill and is located in Suffolk, England. It's famous for its innovative approach to education including a children's democracy based on school meetings. Each pupil has an equal vote along with the teachers. The meetings decide the school rules and deal with those who break them. A guiding principle is "freedom, not licence": coercion is minimized subject to not harming others.

20. Yunus, M., 1998: *Banker to the Poor*, London, Aurum Press. This is the story of the Grameen Bank of Bangladesh, which pioneered microlending and the emancipation of women, succeeding against all expectations.

21. See for instance McNamee, R., 2019: *Zucked: Waking Up to the Facebook Catastrophe*, HarperCollins, also Frenkel, S. and Kang, C., 2021: *An Ugly Truth: Inside Facebook's Battle For Domination*, Bridge Street Press. Roger McNamee is a venture capitalist who helped to launch Facebook,

in its idealistic early days. Since then he has witnessed the political weaponization of Facebook and other social-media platforms, which have used giant behavioural experiments — they might collectively be called a secret 'large hadron collider of experimental psychology'[23] — to build what I'll call artificial intelligences that, unfortunately, aren't yet very intelligent. To be sure, they're clever at exploiting human weaknesses to sell things. But they're downright stupid in other ways, not least in their undermining of democratic social stability and thence, I'll argue, in risking their enterprises' very survival, depending as it does on having a democratic business environment. I'll develop that point near the end my Chapter 3. Central to the problem is the way in which the artificial intelligences have been trained to maximize profits from advertising revenue. They have done so by favouring disinformation and abuse, encouraging it to go viral at lightning speed, often under a cloak of anonymity. If it's emotionally charged, it spreads faster and is more profitable. It's no surprise that this business model, based as it is on disinformation-for-profit and hatred-for-profit or, as journalist Helen Lewis puts it, "the economics of outrage" — whether advertently or inadvertently — has been weaponized by demagogues, conspiracists, extremists, and disinformers of every kind, on a massive scale. See for instance the 2019 book by Nancy L. Rosenblum and Russell Muirhead: *A Lot of People Are Saying: The New Conspiracism and the Assault on Democracy*, Princeton University Press. Regarding the exploitation by demagogues and propagandists — national, international, and multinational — there's a useful short summary by cyber-security expert Clint Watts: "Five generations of online manipulation: the evolution of advanced persistent manipulators," *Eurasia Review*, 13 March 2019. There are also the commercial trollers or 'dragging sites', internet businesses *exclusively* focused on the profits from spreading anonymous lies, abuse, and threats as documented, for instance, by journalist Sali Hughes. Many recent developments in the disinformation and trolling industries — including character assassination using deep-fake artificial intelligence — are documented in the important 2020 book by Nina Schick, *Deep Fakes and the Infocalypse: What You Urgently Need to Know*, Monoray (Octopus Publishing Group). See also the powerful testimony in the 2022 book by journalist and Nobel Peace laureate Maria Ressa, *How to Stand Up to a Dictator*, Penguin, Random House. Strikingly, and aptly, she calls out the disinformation-for-profit basis of the social media — and the power it puts into the hands of demagogues bent on destroying democratic social stability — as "an invisible atom bomb" that has "exploded in our information ecosystem".

22. Pearl, J. and Mackenzie, D., 2018: *The Book of Why: The New Science of Cause and Effect*, London, Penguin. This lucid, powerful, and very readable book describes recently developed forms of Bayesian probability theory[83] and analytics that make causal arrows explicit, clearly distinguishing correlation from causality. Key to this is the probabilistic 'do' operator, or experimentation operator as it might be called. It gives us a precise mathematical framework for dealing with experiments and experimental data. It underpins the most powerful forms of artificial intelligence and big-data analytics. It is applicable to complex problems such as a virus infecting a cell. Its power lies not only behind much of today's cutting-edge open science but also, for instance, behind the covert behavioural experiments done by the social-media platforms.[23]

23. Giant behavioural experiments have been done in secret for many years now by Facebook, and by other social-media platforms, on a scale comparable to that of the large-hadron-collider experiments in particle physics, or perhaps even greater. Experiments using, for example, 'like' buttons can repeatedly test the behaviour of up to billions of subjects. By contrast with the large-hadron-collider experiments — whose data are released into the public domain to advance the science — the giant behavioural experiments have been cloaked in commercial secrecy. The experimental techniques include not only binary button-pressing — press or don't press, and don't stop to think — but also getting users to 'make friends' with digital 'assistants', as they're called, with human-like names, and to play games such as Pokémon Go and its numerous successors. All of this yields vast amounts of data on user behaviour. And from these experiments, using the power of the Bayesian-analytics toolkit,[22] the social-media technocrats have built the artificial intelligences now in use, again under a cloak of commercial secrecy — incorporating not only mathematical models of unconscious human behaviour but also information on the private preferences of billions of individuals. They exploit and manipulate our unconscious assumptions in a precisely targeted way. The sheer scale and power of these

innovations is hard to grasp, along with the political implications[21] but, if only for civilization's sake, it needs to be widely understood.

24. The ozone disinformation campaigns were something that I myself encountered personally, back in the 1980s and 1990s. These well-funded, highly professional campaigns and the longer-running climate disinformation campaigns have been forensically documented in the book by Oreskes, N. and Conway, E. M., 2010: *Merchants of Doubt: How a Handful of Scientists Obscured the Truth on Issues from Tobacco Smoke to Global Warming*, Bloomsbury, 2010. Using formerly secret documents brought to light through anti-tobacco litigation, the authors show how the campaigns were masterminded by the same scientists turned professional disinformers who first honed their skills in the tobacco companies' lung-cancer campaigns.

25. The idea of a cooperative 'symbiosis' between market forces and regulation is hardly new, having been advocated by Adam Smith[40] even if not by all his followers. See also Ref. 2, and the 2020 BBC Reith Lectures by Mark Carney. The idea has been developed further by economist Mariana Mazzucato in her 2018 book *The Value of Everything: Making and Taking in the Global Economy*, Penguin. Chapter 9 suggests how a useful symbiosis can be encouraged when government policy is recognized as "part of the social process which co-shapes and co-creates competitive markets," with innovation and new opportunities among its goals. Mazzucato also argues that such creative developments are being held back by the incoherent popular mythologies, or narratives, that still dominate much of our politics, especially the narratives of possessive individualism and market fundamentalism — the hypercredulous, semi-conscious beliefs that Market Forces are the Answer to Everything and that the only goal should be to maximize individual profits (and never to tax them, nor to regulate anything because, dichotomously speaking, Private is Good and Public is Bad). Another eminent economist, Nobel laureate Paul Krugman, calls these dichotomized narratives "zombie ideas" as a reminder that they should be dead ideas, according to the evidence,[41] but are kept alive artificially — by disinformation paid for by the tiny plutocratic élites that currently take most of the profits.

26. Sulston, J., and Ferry, G., 2003: *The Common Thread: A Story of Science, Politics, Ethics and the Human Genome*, Corgi edn. This important book

by John Sulston and Georgina Ferry gives a first-hand account of how the scientific ideal and ethic prevailed against corporate might — powerful commercial interests that came close to monopolizing the genomic data for private profit. By shutting off open access to the data, the monopoly would have done incalculable long-term damage to biological and medical research, including research on COVID-19 and future pandemics.

27. Strunk, W., and White, E. B., 1979: *The Elements of Style*, 3rd edition, Macmillan. I love the passage addressed to practitioners of the literary arts, whose ambitions transcend mere lucidity: "Even to a writer who is being intentionally obscure or wild of tongue we can say, 'Be obscure clearly! Be wild of tongue in a way we can understand!'"

28. See the article headed "elegant variation" in Fowler, H. W., 1983: *A Dictionary of Modern English Usage*, 2nd edn., revised by Ernest Gowers, Oxford University Press. Fowler's lucid and incisive article is not to be confused with the vague and muddy article under the same heading in the so-called *New Fowler's* of 1996, written by a different author.

29. My first two examples, "If you are serious, then I'll be serious" and "If you are serious, then I'll be also", roughly translate to

你若当真, 我便当真。　你若当真, 我也会。

I don't myself know any of the Chinese languages, but even I can see that the first sentence has an invariant element while the second does not. My Chinese colleagues assure me that the first sentence is clearer and stronger than the second. Lucid repetition works!

30. Littlewood, J. E., 1953: *A Mathematician's Miscellany*, London, Methuen. Republished with extra material by B. Bollobás in 1986, as *Littlewood's Miscellany*, Cambridge University Press.

31. The so-called 'slow manifold' is a multidimensional geometrical object that has a role in the mathematics of fluid-dynamical theory. It's important for understanding the dynamics of jet streams, ocean currents, eddies, and wave–vortex interactions, using an invariant quantity called 'potential vorticity'. In a slightly different sense, it's also important in other branches of fluid dynamics, such as understanding the flow around an aircraft and how the aircraft stays up. It's an object found within the space of all possible fluid flows. It's like something hairy, or fuzzy. More precisely, it's a fractal object. By contrast, a manifold is smooth and non-fractal, like

something bald. I've tried hard to persuade my fluid-dynamical colleagues to switch to 'slow *quasi*manifold', but with scant success. For practical purposes the thing often behaves *as if* it were a manifold (hence *quasi*) even though it isn't one. It's 'thinly hairy'.

32. Medawar, P. B., 1960: *The Future of Man: The BBC Reith Lectures 1959*, Methuen. Near the start of Lecture 2, Medawar remarks that scientists are sometimes said to give the meanings of words "a new precision and refinement", which would be fair enough, he says, were it not for a tendency then to believe that there's such a thing as *the* unique meaning of a word — *the* "true" or "pure" meaning — as if context didn't matter. Then comes his remark that "The innocent belief that words have an essential or inward meaning can lead to an appalling confusion and waste of time", something that very much resonates with my own experience in thesis correcting, peer reviewing, and scholarly journal editing.

33. Hunt, M., 1993: *The Story of Psychology*, Doubleday, Anchor Books. The remarks about the Three Mile Island control panels and their colour coding are on page 606. There were further design flaws, such as making warning indicators inconspicuous and even making parts of the control panels too high to be read by operators.

34. McIntyre, M. E., 1997: "Lucidity and science, I: Writing skills and the pattern perception hypothesis," *Interdisciplinary Science Reviews* **22**, 199–216.

35. See for instance Pomerantsev, P., 2015: *Nothing is True And Everything is Possible — Adventures in Modern Russia*, Faber & Faber. The title incorporates the postmodernist tenet that there's just one Absolute Truth, namely that nothing is true except the belief that nothing is true. (How profound! How ineffable! How Derridian!) Peter Pomerantsev worked as a television producer in Moscow with programme makers in the state's television monopolies, during the first decade of the 2000s. He presents an inside view of how at that time the state controlled the programming through an extraordinarily skilful propagandist, Vladislav Surkov, sometimes called 'The Puppet Master'. Surkov's genius lay in exploiting postmodernist ideas to build an illusion of democratic pluralism in Russia at that time, with mercurial shape-shifting and the skilful and incessant telling of lies as the predominant political weapons — promoting endless confusion — alongside the pressing of binary buttons

to amplify social divisions when expedient. Some Western politicians and campaign advisers have followed the same path, developing what's been euphemistically called a 'post-truth' culture of multifarious, ever-changing 'alternative facts', central to information warfare and what's now called the 'infocalypse', now threatening democracy itself. See the last two books cited in Ref. 21. Postmodernism, as expounded by philosophers such as Jacques Derrida, has provided not only a fertile source of ideas but also a veneer of intellectual respectability, over what would otherwise be described as ordinary mendacious propaganda taken to new extremes.

36. McGilchrist, I., 2009: *The Master and his Emissary: the Divided Brain and the Making of the Western World*, Yale University Press. Iain McGilchrist has worked both as a literary scholar and as a psychiatrist. In this vast and thoughtful book, 'world' often means the *perceived world* consisting of the brain's unconscious internal models, to be discussed in my Chapter 4. In most people, the Master is the right hemisphere with its holistic, integrated 'world', open to multiple viewpoints. The Emissary is the left hemisphere with its dissected, atomistic and fragmented 'world' and its ability to speak and thereby, all too often, to dominate proceedings with its strongly held mindsets and unconscious assumptions.

37. Ramachandran, V. S. and Blakeslee, S., 1998: *Phantoms in the Brain: Human Nature and the Architecture of the Mind*, London, Fourth Estate. Ramachandran is a neuroscientist known for his ingenious behavioural experiments and their interpretation. The phantoms are the brain's unconscious internal models that mediate perception and understanding, to be discussed in my Chapter 4, and this book provides one of the most detailed and penetrating discussions I've seen of the nature and workings of those models and of the roles of the left and right brain hemispheres. Page 59 tells how to produce the 'phantom nose illusion' mentioned in my Chapter 4.

38. The quote can be found in an essay by Max Born's son Gustav Born, 2002: "The wide-ranging family history of Max Born," *Notes and Records of the Royal Society* (London), **56**, 219–262 and *Corrigendum* **56**, 403. Max Born was awarded the Nobel Prize in physics, belatedly in 1954; and the quotation comes from a lecture he gave at a meeting of Nobel laureates in 1964, at Lindau on Lake Constance. The lecture was titled *Symbol and Reality (Symbol und Wirklichkeit)*.

39. Conway, F. and Siegelman, J., 1978: *Snapping: America's Epidemic of Sudden Personality Change*, New York, Lippincott. This sociological study focusing on the American fundamentalist cults of the 1970s remains highly relevant in today's world. Chapter 4 discusses the generic aspects in a thoughtful and interesting way. It records personal accounts of intense religious experience, and similar emotional experiences, going on to take note of their political and commercial exploitation. Musical experiences are included. An ex-preacher, Marjoe Gortner, describes how the techniques of bringing a crowd to religious ecstasy parallel the techniques of a rock concert. Gortner eventually abandoned his preaching career because "as you start moving into the operation of the thing, you get into controlling people and power and money."

40. Tribe, K., 2008: "'Das Adam Smith Problem' and the origins of modern Smith scholarship," *History of European Ideas* **34**, 514–525. This paper provides a forensic overview of Adam Smith's writings and of the many subsequent misunderstandings of them that accumulated in the German, French, and English academic literature of the following centuries — albeit clarified as improved editions, translations, and commentaries became available. Smith viewed the problems of ethics, economics, politics, and human nature together, as a whole, and from more than one angle, and saw his two great works, *The Theory of Moral Sentiments* (1759) and *An Inquiry into the Nature and Causes of the Wealth of Nations* (1776), as complementing each other. He stood by both of them throughout his life. Yes, market forces are useful, but only in symbiosis with written and unwritten regulation,[25] including personal ethics.

41. See for instance Krugman, P., 2020: *Arguing with Zombies: Economics, Politics, and the Fight for a Better Future*, Norton. Also Lakoff, G., 2014: *Don't Think of an Elephant: Know Your Values and Frame the Debate*, Vermont, Chelsea Green Publishing. Economist Paul Krugman and cognitive linguist George Lakoff show how the plutocracies exploiting market fundamentalism[25] perpetuate themselves using their financial power to buy expertise in psychology and persuasion — expertise in the way perception works — exploiting lucidity principles and pressing our binary buttons in the service of disinformation. One of the techniques is to 'frame' things as binary, for instance, setting up a 'debate' to decide between brutal socialism, on the one hand, and free markets, on the other,

as if they were the only two possibilities. The same goes for climate 'versus' the economy, or COVID 'versus' the economy. Dichotomization makes us stupid, as the disinformers well know. The book by Kahneman[6] provides a more general, in-depth discussion of framing, and of related techniques such as anchoring and priming.

42. See for instance Skippington, E., and Ragan, M. A., 2011: "Lateral genetic transfer and the construction of genetic exchange communities," *FEMS Microbiol Rev.* **35**, 707–735. This review article opens with the sentence "It has long been known that phenotypic features can be transmitted between unrelated strains of bacteria." The article goes on to show among other things how "antibiotic resistance and other adaptive traits can spread rapidly, particularly by conjugative plasmids." Plasmids are small packages of DNA, and 'conjugative' means that the plasmid is passed directly from one bacterium to another through a tiny tube called a pilus — even if the two bacteria belong to different species.

43. Wilson, D. S., 2015: *Does Altruism Exist? Culture, Genes, and the Welfare of Others*, Yale University Press. This book by evolutionary biologist David Sloan Wilson focuses on altruism as instinctive behaviour, and on ways in which it can arise from multi-level natural selection in heterogeneous populations. Chapter 2 presents clear and well-verified examples of multi-level selection from laboratory experiments. The examples include the evolution of heterogeneous populations of insects, and of bacteriophage viruses, that exhibited mixtures of selfish and altruistic behaviour. Group-level selection is crucial, alongside individual-level selection, to explaining what was observed in each case. Elsewhere in the book, human belief systems and their consequences are considered. Chapter 1 examines the work of Nobel laureate Elinor Ostrom on belief systems that avoid the 'tragedy of the commons'.[54] Chapters 6 and 7 examine fundamentalist belief systems and their characteristic dichotomizations. Chapter 6 focuses on religious systems including that of the Hutterites of North America, and Chapter 7 on atheist systems including Ayn Rand's famous version of market fundamentalism, at variance with Adam Smith's ideas.[40] Rand's credo, called 'objectivism', and its relation to Friedrich Nietzsche's *Übermensch*, are discussed at some length in Chapter 2 of the 2018 book by the philosopher John Gray, *Seven Types of Atheism*, Allen Lane. See also note 79 below.

44. Wills, C., 1993: *The Runaway Brain: The Evolution of Human Uniqueness*, BasicBooks and HarperCollins. This cogent synthesis by Professor Christopher Wills was years ahead of its time. Near the end of Chapter 12 Wills notes how our ability to imagine fictitious events and "to tell stories that never happened" must have been selected for by "the power, both secular and religious," that it gave to the most skilful storytellers and social manipulators. And, in order to describe the evolutionary processes that gave rise to that ability, Wills builds on his intimate knowledge of genes, palaeoanthropology, and population genetics. The starting point is the quote from Phillip Tobias[4] suggesting the crucial role of multi-timescale genome–culture feedback over millions of years. The author goes on to offer many far-reaching insights, not only into the science itself but also into its turbulent history and controversies. Sewall Wright's discovery of genetic drift is lucidly described in Chapter 8 — that was one of the successes of the old population-genetics models — along with Motoo Kimura's work showing the importance of neutral mutations, now cross-checked at molecular level.[18] As well as advancing our fundamental understanding of evolutionary dynamics, Kimura's work led to the discovery of the molecular-genetic 'clocks' now used to estimate, from genomic sequencing data, the rates of genomic evolution and the times of past genetic bottlenecks. Wills describes how, prior to those developments, progress was held up by dichotomized disputes about neutral mutations *versus* adaptive mutations, failing to recognize that both are important.

45. Rose, H. and Rose, S. (eds.), 2000: *Alas, Poor Darwin: Arguments against Evolutionary Psychology*, London, Jonathan Cape. This thoughtful compendium offers a variety of perspectives on the extreme reductionism or so-called Darwinian fundamentalism of recent decades, what I'm calling simplistic evolutionary theory — as distinct from Charles Darwin's own more complete, more pluralistic view. See also Chapter 10 by an eminent expert on animal behaviour, the late Patrick Bateson FRS, on the word 'instinct' and its controversial technical meanings.

46. Dunbar, R. I. M., 2003: "The social brain: mind, language, and society in evolutionary perspective," *Annual Review of Anthropology* **32**, 163–181. This review offers important insights into the group-level selective pressures on our ancestors, drawing on the primatological, palaeoanthropological, and palaeoarchaeological evidence. See especially the discussion on

pages 172–179. Data are summarized that reflect the growth of brain size and neocortex size over the past three million years, including its extraordinary acceleration beginning half a million years ago, as illustrated in Figure 2 above — by which time, as suggested on Dunbar's page 175, "language, at least in some form, would have had to have evolved", in part to expand the size of the grooming cliques or friendship circles within larger social groups. Even a rudimentary level of language, as with "I love you, I love you", would have been enough to expand such circles, by verbally grooming several individuals at once. The subsequent brain-size acceleration corresponds to what Wills[44] calls runaway brain evolution, under growing selective pressures for ever-increasing linguistic and societal sophistication and group size, culminating — perhaps around the time of the incipient Upper Palaeolithic a hundred millennia ago — in our species' ability to tell elaborate fictional as well as factual stories.

47. Thierry, B., 2005: "Integrating proximate and ultimate causation: just one more go!" *Current Science* **89**, 1180–1183. A thoughtful commentary on the history of biological thinking, in particular tracing the tendency to neglect multi-timescale processes. The fast and slow mechanisms called "proximate causes" and "ultimate causes" are assumed to have no interaction solely *because* "they belong to different time scales" (p. 1182a). "Proximate" and "ultimate" refer respectively to individual-organism and genomic timescales.

48. Rossano, M. J., 2009: "The African Interregnum: the 'where,' 'when,' and 'why' of the evolution of religion," in Voland, E., Schiefenhövel, W. (eds.), *The Biological Evolution of Religious Mind and Behaviour*, Springer-Verlag, pp. 127–141. The African Interregnum refers to the time between the apparent failure of our ancestors' first migration out of Africa, roughly 100 millennia ago, and the second such migration a few tens of millennia later. Rossano's brief but penetrating survey argues that the emergence of belief systems and storytelling having an imagined "supernatural layer" boosted the size, sophistication, adaptability, and hence competitiveness of human groups. As regards the Lake Toba eruption around 70 millennia ago, the extent to which it caused a human genetic bottleneck is controversial, but not the severity of the disturbance to the climate system, like a multi-year nuclear winter. The resulting resource depletion must have severely stress-tested our ancestors' adaptability — giving large, tightly-knit, and socially

sophisticated groups an important advantage. Survival strategies may have included wheeling and dealing between groups. In Rossano's words, the groups were "*collectively* more fit and this made all the difference."

49. Laland, K., Odling-Smee, J., and Myles, S., 2010: "How culture shaped the human genome: bringing genetics and the human sciences together," *Nature Reviews: Genetics* **11**, 137–148. This review notes the likely importance, in genome–culture co-evolution, of more than one timescale. It draws on several lines of evidence. The evidence includes data on genomic sequences, showing the range of gene variants (alleles) in different sub-populations. As the authors put it, in the standard mathematical-modelling terminology, "... cultural selection pressures may frequently arise and cease to exist faster than the time required for the fixation of the associated beneficial allele(s). In this case, culture may drive alleles only to intermediate frequency, generating an abundance of partial selective sweeps... adaptations over the past 70,000 years may be primarily the result of partial selective sweeps at many loci" — that is, locations within the genome. Partial selective sweeps are patterns of genomic change responding to selective pressures yet retaining some genetic diversity, hence potential for future versatility and adaptability. The authors confine attention to very recent co-evolution, for which the direct lines of evidence are now strong in some cases — leaving aside the earlier co-evolution of, for instance, proto-language and its genetically enabled automata.[59] There, we can expect multi-timescale coupled dynamics over a far wider range of timescales, for which direct evidence is much harder to obtain, as discussed also in Ref. 50.

50. Richerson, P. J., Boyd, R., and Henrich, J., 2010: "Gene–culture coevolution in the age of genomics," *Proceedings of the National Academy of Sciences* **107**, 8985–8992. This review takes up the scientific story as it has developed after Wills's book,[44] and usefully complementing Ref. 49. The discussion comes close to recognizing two-way, multi-timescale dynamical coupling but doesn't quite break free of asking whether culture is "the leading *rather than* the lagging variable" in the co-evolutionary system (my italics, to emphasize the false dichotomization).

51. Laland, K., Sterelny, K., Odling-Smee, J., Hoppitt, W., and Uller, T., 2011: "Cause and effect in biology revisited: is Mayr's proximate–ultimate dichotomy still useful?" *Science* **334**, 1512–1516. The dichotomy, between "proximate causation" around individual organisms and "ultimate

causation" on evolutionary timescales, entails a belief that the fast and slow mechanisms are dynamically independent. This review argues that they are not independent, even though the dichotomy is still taken by many biologists to be unassailable. The review also emphasizes that the interactions between the fast and slow mechanisms are often two-way interactions, or feedbacks, labelling them as "reciprocal causation" and citing many lines of supporting evidence. This recognition of multi-timescale feedbacks is part of what's now called the "extended evolutionary synthesis". See also Refs. 47 and 53.

52. See for instance Schonmann, R. H., Vicente, R., and Caticha, N., 2013: "Altruism can proliferate through population viscosity despite high random gene flow," *Public Library of Science, PLoS One* **8**(8), e72043. Improvements in model sophistication, and a willingness to view a problem from more than one angle, show that group-selective pressures can be effective.

53. Danchin, E. and Pocheville, A., 2014: "Inheritance is where physiology meets evolution." *Journal of Physiology* **592**, 2307–2317. This complex but very interesting review is one of two that I've seen — the other being Ref. 51 — that go beyond Refs. 49 and 50 in recognizing the importance of multi-timescale dynamical processes in biological evolution. There has been a widespread assumption, perhaps unconscious, that timescale separation implies dynamical decoupling (see also Ref. 47). In reality there is strong dynamical coupling, the authors show, involving an intricate interplay between different timescales. It's mediated in a rich variety of ways including not only niche construction and genome–culture co-evolution but also, at the physiological level, developmental plasticity along with the non-genomic heritability now called epigenetic heritability. One consequence is the creation of hitherto unrecognized sources of heritable variability, the crucial raw material on which natural selection depends for its effectiveness. See also the *Nature* Commentary by Laland, K. *et al.*, 2014: "Does evolutionary theory need a rethink?" *Nature* **514**, 161–164. (In the Commentary, for 'gene' read 'replicator' including regulatory DNA.)

54. Werfel, J., Ingber, D. E., and Bar-Yam, Y., 2015: "Programed death is favored by natural selection in spatial systems," *Physical Review Letters* **114**, 238103. This detailed modelling study illustrates yet again how various 'altruistic' traits are often selected for, in models that include population heterogeneity and group-level selection. The paper focuses on the ultimate

unconscious altruism, *mortality* — the finite lifespans of most organisms. Finite lifespan is robustly selected for, across a wide range of model assumptions, simply because excessive lifespan is a form of selfishness leading to local resource depletion. The tragedy of the commons, in other words, is as ancient as life itself. The authors leave unsaid the implications for our own species.

55. Contera, S., 2019: *Nano comes to life: How Nanotechnology is Transforming Medicine and the Future of Biology*, Princeton University Press. Sonia Contera shows how cutting-edge biological and medical research are breaking free from the mindsets of simplistic evolutionary theory, with its "reductionist vision" that genes govern everything — its view of organisms as "mere biochemical computers executing a program... encoded in genes". She points out that "current medical bottlenecks will remain blocked" until medical science completes its escape from the 'blueprint' or 'genes govern everything' mindset. She also describes amazing new experiments to explore artificial self-organizing or self-assembling nanometre-scale structures, and functional machines, made of DNA or proteins. (A nanometre is a millionth of a millimetre.) These self-assemble just as their natural counterparts do, from the most simple to the most complex including ordinary crystals, viruses, and multicellular organisms.[57] Progress on the protein folding, or protein self-organization, problem prior to the AlphaFold breakthrough is discussed under the heading *Protein Nanotechnology*.

56. Pinker, S., 1997: *How the Mind Works*. London, Allen Lane. The author invokes "mathematical proofs from population genetics" in support of what amounts to simplistic evolutionary theory (Chapter 3, page 163, section on "Life's Designer"). The author is silent on which population-genetics equations were used in these "proofs". However, as the book proceeds, it becomes clear that the author is referring to the equations of the old population-genetics models, in their simplest versions that do not *prove* simplistic evolutionary theory but, rather, *assume* it in writing down the equations. In particular, the models exclude group-level selection by confining attention to averages over whole populations, conceived of as statistically homogeneous and as living in a fixed, homogeneous environment. Notice the telltale phrase "on average" in the Section "I and Thou" in Chapter 6, on page 398. Not even the famous Price equation, perhaps the first attempt to allow for population heterogeneity, is mentioned,

nor multiple timescales, nor Monod's arguments.[3] Almost exactly the same critique can be made of Richard Dawkins's famous book *The Selfish Gene*, which makes passionate assertions to the effect that simplistic evolutionary theory and its game-theoretic aspects, such as reciprocal altruism, have been rigorously and finally established by mathematical analysis — again meaning the old population-genetics models — and that any other view is wrong or muddled. See also Ref. 57.

57. Dawkins, R., 2009: *The Greatest Show On Earth*. London, Bantam Press. I am citing this book alongside Ref. 56 for two reasons. First, Chapter 8 beautifully illustrates why emergent properties and self-assembling building blocks (automata) are such crucial ideas in biology. It becomes clear why the genetic-blueprint idea is such a gross oversimplification. Chapter 8 illustrates the point with a sequence of examples beginning with the self-assembly[55] of viruses and ending with the Nobel-prize-winning work of my late friend John Sulston FRS, on the self-assembly of an adult nematode worm, *Caenorhabditis elegans*, from its embryo. All this gets us significantly beyond simplistic evolutionary theory, as John suggested I call it. Second, however, the book persists with the mindset against group-level selection. Any other view is, it says, an amateurish fallacy (end of long footnote in Chapter 3, page 62). No hint is given as to the basis for this harsh verdict; but as in *The Selfish Gene* the basis seems to be an unquestioning faith in particular sets of mathematical equations, namely those defining the old population-genetics models. Those equations exclude group-level selection by assumption, or include at most only weak forms of it. And Jacques Monod, who argued that group-level selection was critical to our ancestors' evolution,[3] was hardly an amateur. He was a great scientist and a very sharp thinker who, as it happens, was also a Nobel laureate.

58. Segerstråle, U., 2000: *Defenders of the Truth: The Battle for Science in the Sociobiology Debate and Beyond*, Oxford University Press. This important book gives insight into the disputes about natural selection over past decades. It's striking how dichotomization kept muddying those disputes, even among serious and respected scientists. There were misplaced pressures for parsimony of explanation, forgetting Einstein's famous warning not to push Occam's Razor too far. Again and again the disputants felt, it seems, that 'we are right and they are wrong' and that there's only one truth, to be viewed in only one way. Again and again, progress was impeded

by a failure to recognize complexity, multidirectional causality, different levels of description, and multi-timescale dynamics. And the confusion was sometimes made even worse, it seems, by failures to disentangle science from raw politics.

59. The Acheulean toolmaking skills call for great accuracy and delicacy in striking flakes off brittle stones such as obsidian or flint. And the point about language is not that it could have been used to *describe* the skills. Even today one cannot acquire such skills just by hearing them described. Rather, language would have operated on the level of personal relations. Even a rudimentary language ability, with a rudimentary syntax, would have helped a teacher to encourage a novice in efforts to acquire the skills. "Try again. Hit it there! That's better!" Straight away that's a selective pressure to develop language beyond the simple I-love-you, you-tickle-me social grooming level.[3, 46] And with it there's a pressure to develop future thinking — enabling the novice to imagine becoming a good toolmaker one day. "Keep at it. You'll get there!"[60] I heard these points made in a BBC radio interview with an expert on the Oldowan and Acheulean stone tools, Dietrich Stout. Professor Stout and co-workers have done interesting work rediscovering the toolmaking skills and tracing their imprint upon brain scans, including neural pathways converging on Broca's area in the brain, crucial to speech generation.

60. On future thinking — clearly helpful in a teaching situation — there's evidence that, as primatologist Jane Goodall put it, "Chimpanzees can plan ahead, too, at least as regards the immediate future. This, in fact, is well illustrated at Gombe, during the termiting season: often an individual prepares a tool for use on a termite mound that is several hundred yards away and absolutely out of sight." The quote is from Chapter 2 of the famous 1990 book by Goodall, *Through a Window: My Thirty Years with the Chimpanzees of Gombe*, Weidenfeld & Nicholson and Houghton Mifflin. Recent field work by primatologist Catherine (Cat) Hobaiter and co-workers has revealed many more of the cognitive abilities of the great apes, as observed in the wild, including gestural proto-language. Chimpanzees, bonobos, and other great apes use intentional communication via gestures, with 'vocabularies' of many tens of distinct gestures. Some of the gestures resemble those of human infants. And it's well known that chimpanzees

and bonobos understand rudimentary syntax, as in distinguishing 'Me tickle you' from 'You tickle me.'

61. Aiello, L. C., 1996: "Terrestriality, bipedalism and the origin of language." *Proceedings of the British Academy* **88**, 269–289. Reprinted in: Runciman, W. G., Maynard Smith, J., and Dunbar, R. I. M. (eds.), 1996: *Evolution of social behaviour patterns in primates and man*, Oxford University Press, and British Academy. The paper presents a closely argued discussion of brain-size changes in our ancestors over the past several million years, alongside several other lines of palaeoanatomical evidence. Taken together, these lines of evidence suggest an "evolutionary arms race" between competing tribes whose final stages led to the full complexity of language as we know it. Such a picture is consistent with Monod's and Wills's arguments,[3, 44] the final stages corresponding to what Wills called runaway brain evolution. Aiello cites evidence for a date "earlier than 100,000 years" for the first occurrence of art objects in the archaeological record, suggesting that language was highly developed by then, setting the stage for the Upper Palaeolithic.

62. Harari, Y. N., 2014: *Sapiens: A Brief History of Humankind*. Yuval Noah Harari's famous book explores aspects of our ancestors' evolution, referring to the palaeogenetic, palaeoarchaeological, and archaeological records and to various attempts at explanatory theories, while being careful to note where evidence is lacking. The book puts emphasis on what's called the 'cognitive revolution' near the start of the Upper Palaeolithic. That's a late stage in the runaway brain evolution discussed by Wills[44] and also, for instance, by Rossano;[48] see Figure 2 above. As Wills and Rossano also point out, the most crucial development was probably the ability to imagine nonexistent worlds and to tell stories about such worlds, promoting group solidarity and versatility in very large and powerful groups or tribes. Of course, a prerequisite would have been a language ability already well developed.

63. See for instance the short paper by Montgomery, S., 2018: "Hominin brain evolution: the only way is up?" *Current Biology* **28**, R784–R802. Montgomery's Figure 1b is an updated version of Figure 2 above, with data from more fossil finds and improved measurement accuracy. The overall pattern is the same, apart from two recent new finds, *Homo floresiensis* and *Homo naledi*. Regarding recent discoveries about gene flow across different

strands of our ancestry, the Wikipedia article "Interbreeding between archaic and modern humans" gives a useful summary, referring to events several tens of millennia ago for which DNA evidence is now available.

64. Kegl, J., Senghas, A., and Coppola, M., 1999: "Creation through contact: sign language emergence and sign language change in Nicaragua," in *Language Creation and Language Change: Creolization, Diachrony, and Development*, edited by Michel DeGraff, MIT Press, pp. 179–237. This is a detailed compilation of the main lines of evidence. Included are studies of the children's sign-language descriptions of videos they watched. Also, there are careful discussions of the controversies among linguists, including those who cannot accept the possibility that genetically enabled automata for language might exist.

65. Pinker, S., 1994: *The Language Instinct*, Allen Lane. The Nicaraguan case is briefly described in Chapter 2, as far as it had progressed by the early 1990s.

66. See for instance Senghas, A., 2010: "The emergence of two functions for spatial devices in Nicaraguan Sign Language," *Human Development* (Karger), **53**, 287–302. This later study uses video techniques as in Ref. 64 to trace the development and standardization, by successive generations of young children, of syntactic devices in the signing space.

67. See for instance Ehrenreich, B., 1997: *Blood Rites: Origins and History of the Passions Of War*, London, Virago and New York, Metropolitan Books. Barbara Ehrenreich's insightful and penetrating discussion contains much wisdom, it seems to me, not only about war but also about the nature of mythical deities and about human sacrifice, ecstatic suicide, and so on — echoing Stravinsky's *Rite of Spring* and long pre-dating Nine-Eleven and ISIS/Daish. (Talk about ignorance being expensive!)

68. Lüthi, D., Le Floch, M., Bereiter, B., Blunier, T., Barnola, J.-M., Siegenthaler, U., Raynaud, D., Jouzel, J., Fischer, H., Kawamura, K., and Stocker, T. F., 2008: "High-resolution carbon dioxide concentration record 650,000–800,000 years before present," *Nature* **453**, 379–382.

69. The abruptness of the Dansgaard–Oeschger warmings and their possible mechanisms are discussed in my 2022 paper "Climate uncertainties: a personal view," *Meteorology* **1**, 162–170. It's a short survey discussing various tipping points and bringing together key evidence from the literature, most notably Alley, R. B., 2000: "Ice-core evidence of abrupt climate changes," *Proceedings of the National Academy of Sciences* **97**,

1331–1334. Richard Alley is a respected expert. His paper summarizes the evidence from Greenland ice cores. That evidence is key because annual layers in the ice can be distinguished and counted, allowing changes taking only a few years to be clearly seen and precisely dated. The astonishingly short timescales of the Dansgaard–Oeschger warmings include those of the most recent or 'zeroth' such warming, about 11.5 millennia ago, which Alley describes as taking place in three 5-year steps within a 40-year period. See also Ref. 71.

70. Alley, R. B., 2007: "Wally was right: predictive ability of the North Atlantic 'conveyor belt' hypothesis for abrupt climate change," *Annual Review of Earth and Planetary Sciences* **35**, 241–272. This paper incorporates a very readable, useful, and informative survey of the relevant palaeoclimatic records, and recent thinking about them. Wally Broecker's famous 'conveyor belt' is a metaphor for the oceans' global-scale overturning circulation, an important part of which is the Atlantic overturning circulation. The metaphor has greatly helped efforts to understand the variability observed during the glacial cycles. My own assessment is, however, that Wally was *partly* right. Despite its usefulness, the metaphor embodies a fluid-dynamically unrealistic assumption, namely that shutting off North Atlantic deep-water formation also shuts off the global-scale return flow. If you jam a real conveyor belt somewhere, then the rest of it stops, too. In this respect, the metaphor needs refinements such as those argued for by Trond Dokken and co-workers,[71] recognizing that parts of the 'conveyor' can shut down while other parts continue to move and transport heat and salt at significant rates. As they point out, such refinements are likely to be important for understanding the abruptness of the Dansgaard–Oeschger warmings, and a similar tipping point that might soon occur in the Arctic Ocean.

71. Dokken, T. M., Nisancioglu, K. H., Li, C., Battisti, D. S., and Kissel, C., 2013: "Dansgaard–Oeschger cycles: interactions between ocean and sea ice intrinsic to the Nordic Seas," *Paleoceanography* **28**, 491–502. This is the first fluid-dynamically credible explanation that I've seen of the extreme rapidity — the abruptness — and the large magnitudes of the Dansgaard–Oeschger warming events. Those events left clear imprints in ice-core and sedimentary records all over the Northern Hemisphere and were so sudden, and so large in magnitude, that a tipping-point mechanism must have been

involved. The proposed explanation is evidence-based and represents the only mechanism suggested, so far, that can act quickly enough — involving the sudden disappearance of sea ice covering the Nordic Seas when an intrusion of warmer subsurface Atlantic water became buoyant enough to break through to the surface. Figure 4 above is Figure 2a of this paper, whose full text is open-access and available at https://doi.org/10.1002/palo.20042. The authors point out that today's Arctic sea ice might be vulnerable to the same tipping-point mechanism.

72. Bail, C., 2021: *Breaking the Social Media Prism: How to Make Our Platforms Less Polarizing*, Princeton University Press. This important book by Professor Chris Bail, a sociologist at Duke University, describes recent experimental-psychological studies aimed at deepening our understanding of the social media and their political effects. The 'prism' is a metaphor to suggest the distorted vision via social media that highlights the extremist views of small minorities, giving them undue prominence because it's profitable. Professor Bail points out that tackling the problem calls for ideas far more sophisticated than the simplistic idea of breaking open the echo chambers. The last chapter suggests possible ways forward based on the experimental findings.

73. One of these experimental studies demonstrated the flexibility of group-identity feelings among Liverpool and Manchester United football fans. In an extended interview published in *Social Science Space*, https://www.socialsciencespace.com/2016/02/stephen-reicher-on-crowd-psychology/, Professor Reicher summarized the experiments thus: "I was very interested in this idea of shared identity and helping, so we did some experimental studies with a colleague, Mark Levine... We did a series of studies on categories and helping. Very simple study: You get people who are Manchester United fans, and you talk to them as Manchester United fans, and you say well, look, we are going to do a study in another building, and as they walk along to the other building, somebody runs along, falls over, hurts themselves, wearing either a Manchester United shirt, a Liverpool shirt or a red t-shirt, and they help the person in the Manchester United shirt, not the Liverpool shirt and not the red t-shirt... but the really interesting thing... was a different condition where again, we take... Manchester United fans, but this time we address them as football fans, we say we are doing research on football fans, they go to the other building, somebody runs along, falls over

wearing a Manchester United shirt, a Liverpool shirt or a red t-shirt. This time... they help the Manchester United fan, *they help the Liverpool fan,*" (my italics) "and they don't help the person in the red t-shirt... evidence... showing how varying the ways in which we define identities, varies the limits of solidarity." More recently Professor Clifford Stott, an ex-student of Reicher's, has worked with police to demonstrate how rioting in a crowd can be reduced or prevented, through understanding the dynamics of group identities in their response to different policing methods.

Another striking demonstration of flexible group identity was an electoral success in Turkey in June 2019, in which political traction came not from demagoguery and binary-button-pressing but rather, to some people's surprise, from a politicians' colourful handbook called the *Book of Radical Love*, switching the focus away from antagonism, and toward pluralistic core values and "caring for each other" — even for one's political opponents! Arguably, that success was yet another demonstration of caring as a deeply unconscious, deeply instinctive, powerful part of human nature, adding to the examples in Refs. 2, 10, and 11.

74. The point that perceived times differ from physical times is not only simple but also easy to check experimentally as I'll show. But it's hard to find in the philosophical literature. Indeed, some professional philosophers such as Hugh Mellor have explicitly denied it, in his 1981 book *Real Time*. Others have recognized it, for instance Henri-Louis Bergson and Daniel Dennett, in publications dating, respectively, from 1889 and 1991. But each in his own way makes it part of something complicated. See for instance the seventy pages of discussion in Chapters 5 and 6 of Dennett's 1991 book *Consciousness Explained*, Penguin. Among other things, the discussion gets immersed in the usual philosophical quagmire surrounding the idea of 'free will' and the timing of decisions to act.[75] A quotation from Mellor's book can be found on Dennett's page 149, and the famous decisions-to-act experiment by Grey Walter is described on page 167.

75. The term *acausality illusion* does not seem to be in general use, but I think it helps to underline the simplicity of something often regarded as complicated and mysterious.[74] See McIntyre, M. E., 1997: "Lucidity and science, II: From acausality illusions and free will to final theories, mathematics, and music," *Interdisciplinary Science Reviews* **22**, 285–303. On the issue of free will I point out that, contrary to what's sometimes asserted in the philosophical

literature, decisions-to-act experiments like those of Benjamin Libet and Grey Walter[74] have nothing to say on that issue. Rather, the experiments add to our examples of acausality illusions. Patients were asked to carry out an action such as pressing a button, at a time of their choosing, while the experimenters measured the associated brain activity. The timespan of the brain activity was the usual several hundred milliseconds with, inevitably, some of the activity preceding the perceived time of willing the action. The philosophical literature tends to miss the point that perceived times are — and can only be — properties of the brain's unconscious internal models that mediate perception[76, 77] and not the times of any particular brain-activity events. Of course a perceived time must come from interrogating an internal model already formed, or forming, and so is always — and can only be — something within short-term or longer-term memory as with the "gunfight in slow motion", Tony Kofi's three-storey fall, and for instance Professor David Eagleman's famous SCAD divers.

76. Gregory, R. L., 1970: *The Intelligent Eye*, Weidenfeld and Nicolson. This great classic is still well worth reading. It's replete with beautiful and telling illustrations of how vision works. Included is a rich collection of stereoscopic images viewable with red–green spectacles. The brain's unconscious internal models that mediate visual perception are called "object hypotheses", and the active nature of the process whereby they're selected is clearly recognized, along with the role of prior probabilities. There's a thorough discussion of the standard visual illusions as well as such basics as the *perceptual grouping* studied in Gestalt psychology. In a section on language and language perception, Chomsky's 'deep structure' is identified with the repertoire of unconscious internal models used in decoding sentences, and their functionality. The only points needing revision are speculations that the first fully developed languages arose in very recent millennia[13] and that they depended on the invention of writing. That's now refuted by the evidence summarized in my Chapter 3, showing that there are genetically enabled automata for the deep structure of language, including syntactic function.

77. Hoffman, D. D., 1998: *Visual Intelligence: How We Create What We See*, Norton. This book updates Ref. 76, with further illustrations and some powerful insights into the way visual perception works. Chapter 6 describes a case in which a patient, part of whose visual cortex had been

damaged by a stroke, saw moving objects as a series of snapshots. For instance, when tea was being poured into a cup, "the fluid appeared to be frozen, like a glacier." The stroke damage was to a small visual-cortex area called V5. See also Seth, A., 2021: *Being You: A New Science of Consciousness*, Faber. Neuroscientist Professor Anil Seth aptly calls the perceptual model-fitting process 'controlled hallucination', referring to 'control' by the incoming data. He recognizes the importance of the body image and the self-model.

78. See for instance Gilbert, C. D., and Li, W. 2013: "Top-down influences on visual processing," *Nature Reviews (Neuroscience)* **14**, 350–363. This review presents anatomical and neuronal evidence for the active, prior-probability-dependent nature of perceptual model-fitting. Top-down and bottom-up pathways are also called feedback and feedforward, respectively. Thus, "Top-down influences are conveyed across... descending pathways covering the entire neocortex... The feedforward connections... ascending... For every feedforward connection, there is a reciprocal [descending] feedback connection that carries information about the behavioural context... Even when attending to the same location and receiving an identical stimulus, the tuning of neurons can change according to the perceptual task that is being performed", etc.

79. Beliefs about veridical perception are further discussed in Ref. 75. And, strange though it may seem, veridical perception was part of the gospel according to Ayn Rand.[43] In her famous writings on 'objectivism' she says, in effect, that veridical perception is an Absolute Truth. As the Wikipedia article on objectivism puts it, Rand's claim is that "human beings have direct contact with reality through sense perception... perception, being determined physiologically, is incapable of error... optical illusions are errors in the conceptual identification of what is seen, not errors of sight itself." Notice the eloquence and fluency of language alongside the complete absence of logic-checking. Why should a mechanism be "incapable of error" just because it works "physiologically"?

80. Marr, D. C., 1982: *Vision: a computational investigation into the human representation and processing of visual information*, San Francisco, Freeman. The late David Marr's celebrated classic was a landmark in vision research, focusing on bottom-up visual processing in its earliest stages, such as edge detection and stereoscopic image-matching.

81. For the ocean-eddy problem — which is a crucial part of the climate problem — we can't actively experiment with the real ocean but we can take what are called eddy-resolving simulations of oceanic flows, whose realism can be checked against real oceanic flows, and *experiment with those simulations* to find patterns of behaviour of, for instance, Gulf Stream eddies, using the Bayesian causality theory.[22] Work of this kind is still in its infancy. See for instance Zanna, L. and Bolton, T., 2020: "Data-driven equation discovery of ocean mesoscale closures," *Geophys. Res. Lett.* **47**, e2020GL088376.

82. See for instance Unger, R. M., and Smolin, L., 2015: *The Singular Universe and the Reality of Time: A Proposal in Natural Philosophy*, Cambridge University Press. Roberto Mangabeira Unger and Lee Smolin — two respected thinkers in their fields, philosophy and physics, respectively — present a profound and wide-ranging discussion of how progress might be made in fundamental physics and cosmology despite formidable conceptual difficulties (some of which, incidentally, are reduced by taking the model-fitting view as I advocate it). One of the difficulties is a tendency to conflate the outside world with our mathematical models of it, what Edwin T. Jaynes[83] called the mind-projection fallacy. Thus, for instance, the Platonic world of perfect mathematical forms[84] is seen as something external, as well as immutable and everlasting — as Plato himself seems to have thought — and something that *contains* the Universe we live in. According to this view, as advocated for instance by cosmologist Max Tegmark, the Universe and everything in it are purely mathematical entities, no more and no less. Such a conclusion is arguably implausible (and see my Chapter 6), but in any case, as Unger and Smolin point out, it belongs to metaphysics, rather than to physics or to any other kind of science.

83. Jaynes, E. T., 2003: *Probability Theory: The Logic of Science*, edited by G. Larry Bretthorst, Cambridge University Press. This vast posthumous work by the late Edwin T. Jaynes has a strong focus on the theoretical foundations of probability and statistics in their powerful 'Bayesian' form, underpinned by the theorems of Richard Threlkeld Cox. Those theorems, and the work of pioneers like Adrian Smith FRS in the 1970s and 1980s, were important steps toward today's still more powerful *Bayesian causality theory.*[22.] Much of Jaynes's book digs into the technical detail, but there are instructive journeys into history as well, especially in Chapter 16. For many

decades, progress was held up by dichotomized disputes reminiscent of the disputes over biological evolution[44, 58] — and similarly damaging to progress. A narrow view of statistical inference, called 'frequentist', is useful in some problems but was taken by professional statisticians to be the only permissible view. As with energy budgets and selfish genes, 'usefulness' had morphed into 'Answer to Everything'. And the frequentist view seems to have been entangled with the mind-projection fallacy,[82] masquerading as 'objectivity'. There are mathematical entities called probability distribution functions, and these were seen as things in the outside world rather than as components of mathematical models — the veridical perception fallacy[79] in a new guise. Furthermore, because of dichotomization, the frequentist view was mistakenly seen as excluding the more powerful and versatile Bayesian view, even though we can now see the latter as for the most part *including* the former — once we recognize probabilities as model components and make explicit the information on which they're contingent, in the notation of so-called 'conditional' probabilities. (Explicitness principle again.) A quick introduction to the basics of all this, including Cox's theorems, is given in my 2007 paper "On thinking probabilistically," http://www.damtp. cam.ac.uk/user/mem/index.html#thinking-probabilistically, published in *Extreme Events* (Proceedings of the 15th 'Aha Huliko'a Winter Workshop) edited by P. Müller, C. Garrett, and D. Henderson, SOEST publications, University of Hawaii at Manoa, pp. 153–161.

84. Penrose, R., 1994: *Shadows of the Mind: A Search for the Missing Science of Consciousness,* Oxford University Press. This fascinating book includes a discussion of the Platonic world as seen by a great mathematician, Roger Penrose FRS, who in 2020 received a Nobel prize for his powerful black-hole singularity theorem. (That theorem, a property of the model described by Einstein's equations, was published in 1965 and predicted that black holes can form in a huge range of circumstances, not just the special circumstances previously considered — motivating an early search for them and later the LIGO work described in Ref. 16.) I've dared to disagree, however, with Penrose's take on consciousness and the Platonic, for reasons given in my Chapter 6 and at greater length in an appendix to Ref. 75, headed *On mathematical truth.*

85. See for instance Smythies, J. 2009: "Philosophy, perception, and neuroscience," *Perception* **38**, 638–651. This review documents parts

of what I call the quagmire of philosophical confusion about the way perception works. The discussion begins by noting, among other things, the persistence of the fallacy that perception is what it seems to be subjectively, namely veridical[79] in the sense of being direct and independent of any model-fitting process, a simple mapping between appearance and reality. That's still taken as self-evident, we're told, even by some professional philosophers despite the evidence from experimental psychology and neuroscience, as summarized, for instance, in Refs. 37, 76, 77, and 78. Then a peculiar compromise is advocated, in which perception is *partly* direct, and *partly* works by model-fitting, so that "what we actually see is always a mixture of reality and virtual reality" [*sic*; p. 641]. Of greater interest, though, is a summary of some old clinical evidence, from the 1930s, that gave early insights into the brain's different model components. Patients described their experiences of vision returning after brain injury, implying that different model components recovered at different rates and were detached from one another at first. On pp. 641–642 we read about recovery from a particular injury to the visual regions in the occipital lobe: "The first thing to return is the perception of movement. On looking at a scene the patient sees no objects, but only pure movement... Then luminance is experienced but... formless... a uniform white... Later... colors appear that float about unattached to objects (which are not yet visible as such). Then parts of objects appear — such as the handle of a teacup — that gradually coalesce to form fully constituted... objects, into which the... colors then enter."

86. Here's a little challenge to your powers of observation. After getting out of the bath and pulling the plug, observe the sense in which the water begins to circulate around the plughole. (It can be either way, regardless of whether you live in the Northern or the Southern Hemisphere — depending not on the Earth's rotation but on how you got out of the bath.) If the water circulates in a clockwise vortex, what's often most eye-catching is an appearance of anticlockwise motion. But that's a motion not of the water but of the waves on its surface. The waves you generated when you got out of the bath are refracted by the vortex in such a way as to propagate against the vortex flow.

87. I owe the first part of the aphorism to consultant psychiatrist Mark Salter, whom I met at the Hay-on-Wye festival *How The Light Gets In*.

In 2013, I gave a talk there and took part in a discussion in which the issue of free will came up. Mark provoked us with "Free will is a biologically indispensable illusion." That struck me as a rather neat statement, albeit a touch mysterious. I took it to mean, though, that what's indispensable is the need to regard free will as illusory when trying to understand biological functioning — that is, biological functioning in the usual sense. At the physiological and molecular levels usually considered, as distinct from the mental and social levels, it's necessary to exclude the concept of free will as irrelevant. I can't resist adding a slightly different take on it, at the mental and social levels, from the father of chaos theory Professor Edward N. Lorenz: "We must wholeheartedly believe in free will. If free will is a reality, we shall have made the correct choice. If it is not, we shall still not have made an incorrect choice, because we shall not have made any choice at all, not having a free will to do so." From Lorenz, E. N., 1995: *The Essence of Chaos*, University of Washington Press. Ed Lorenz was one of the most lucid thinkers among my colleagues. He made profound contributions to understanding the chaotic dynamics of weather and climate.[117, 118]

88. Feynman, R. P., Leighton, R. B., and Sands, M., 1964: *Lectures in Physics*, Addison-Wesley. The action integral and Newton's equations are discussed in Chapter 19 of volume II, *Mainly Electromagnetism and Matter*.

89. Platt, P., 1995: "Debussy and the harmonic series," in *Essays in honour of David Evatt Tunley*, ed. Frank Callaway, pp. 35–59, Perth, Callaway International Resource Centre for Music Education, School of Music, University of Western Australia, ISBN 086422409 5. Debussy also exploited, for instance, the powerful musical pattern called the octatonic scale.

90. Besides blue notes there are countless other examples. One can play games with the harmony and with the so-called 'expression' — the artistic treatment of emotional affect — creating or relieving tension as the pitch is slightly pushed up or down. More subtly, fine pitch variations and fluctuations can be perceived as part of the 'tone colour' produced by, for instance, a violin.[93] All these points tend to be missed in theoretical discussions of musical fixed tunings, or 'temperaments'. One sometimes encounters what seems to be an unconscious assumption, at variance with actual performance practice, that non-keyboard musical performances comply rigidly with one or another fixed temperament. One also encounters an assumption that temperaments, including the 'equal temperament' of the

standard keyboard, are evil compromises with the 'pure' or 'just' tunings of low-order harmonic-series subsets — as if the consequences were entirely negative rather than enriching.

91. The theme of the little fugue comes from the telephone number of two dear friends — yes, playing with numbers leads to musical ideas. I wrote it partly for the friends' wedding anniversary celebration and partly in memory of my late father A. K. McIntyre, who was a respected neurophysiologist, and who was much loved as a kind and gentle man. The duration is about five and a half minutes. In the recording presented in Figure 18, Ruth McIntyre plays the piano, Vivian Williams the cello, and I the violin, and the recording and mastering were by Jeffrey Ginn Music Studios. Score and parts are available at http://www.damtp.cam. ac.uk/user/mem/papers/LHCE/music-index.html, under "Fugue on a Telephone Number."

92. Boomsliter, P. C., and Creel, W., 1961: "The long pattern hypothesis in harmony and hearing," *Journal of Music Theory* (Yale School of Music), 5(1), 2–31. This wide-ranging and penetrating discussion was well ahead of its time and is supported by the authors' ingenious psychophysical experiments, which clearly demonstrate waveform-cycle counting as distinct from Fourier analysis. On the purely musical issues, there is only one slight lapse, in which the authors miss the context dependence of tonal major–minor distinctions. On the other hand the authors clearly recognize, for instance, the biological relevance of "efficient listening for sensing the environment" (p. 13), what's now called auditory scene analysis.

93. The interested reader is referred to Figure 3 of McIntyre, M. E., and Woodhouse, J., 1978: "The acoustics of stringed musical instruments," *Interdisciplinary Science Reviews* **3**, 157–173. The figure shows why the intensities of different overtones fluctuate out of step with each other during vibrato. We carried out our own psychophysical experiments to verify the strong effect on perceived tone quality. Further work on the acoustics of vibrating strings, reeds, and air jets is reported in McIntyre, M. E., Schumacher, R. T., and Woodhouse, J., 1983: "On the oscillations of musical instruments," *Journal of the Acoustical Society of America* **74**, 1325–1345, and in Scavone, G., and Smith, J. O., 2021: "A landmark article on nonlinear time-domain modeling in musical acoustics,"

Journal of the Acoustical Society of America **150**(2), R3–R4, https://doi.org/10.1121/10.0005725.

94. Personal communication, 2001–present. Tim was elected FRS in 2003 for his work on the chaotic dynamics of weather and climate. His remarkable new book, Ref. 118, shows not only how chaos theory is fundamental to weather and climate dynamics — see also Ref. 117 — but also how it might resolve the enigmas at the foundations of quantum theory. For the quantum-theoretic issues, see also for instance Palmer, T. N., 2021: "Undecidability, fractal geometry and the unity of physics," a prizewinning essay written for the Foundational Questions Institute Prize Essay competition and published in *Undecidability, Uncomputability, and Unpredictability*, ed. Anthony Aguirre, Zeeya Merali, and David Sloan, Springer-Verlag, pp. 81–95. Ideas about turbulence come in somewhat obliquely, through chaos theory and through a reference to limitingly small scales of turbulent eddies. That aspect of the research is, however, still in its infancy. Some of the technicalities are further discussed in Ref. 118 and in Palmer, T. N., 2020: "Discretisation of the Bloch sphere, fractal invariant sets and Bell's theorem," *Proceedings of the Royal Society* **A 476**, 20190350. Regarding possible chaos at the minuscule Planck lengthscale at which quantum effects mesh with gravity, an aspect *unlike* ordinary turbulence is that spacetime might itself become chaotic, with ordinary space and time as emergent properties belonging only to much larger scales — such as the atomic and subatomic scales measured by the 'Bohr radius' and 'proton charge radius'. See for instance Chapter 5 of Carlo Rovelli's 2015 book *Seven Brief Lessons on Physics*, Penguin.

95. Pierrehumbert, R. T., 2010: *Principles of Planetary Climate*, Cambridge University Press. Professor Raymond Pierrehumbert FRS is a brilliant thinker and a leading expert on the climate dynamics of the Earth throughout its history, and on the climate dynamics of other planets. The physics, chemistry, biology, and other relevant aspects are explained in a masterly way. For instance, Section 8.4 gives the clearest explanation I've seen of how carbon dioxide is stored in the Earth's deep oceans, mostly as bicarbonate ions. Some subtle and tricky chemistry is involved, sensitive to oceanic acidity. On multi-century and multi-millennial timescales, depending on the acidity, some of the carbon can be stored by precipitation

as limestone sludge. On this point see also, for instance, Ref. 100, and on today's climate models see Refs. 117 and 118.

96. See for instance Archer, D., 2009: *The Long Thaw: How Humans Are Changing the Next 100,000 Years of Earth's Climate*, Princeton University Press. The author is an expert on the so-called 'carbon cycle' and its role in the climate system. He considers the system as a whole, with special attention to sea levels and to what happens to the carbon dioxide we're injecting, and to the natural mechanisms for recovery. The natural recovery timescales stretch out beyond a hundred millennia, as is clearly evidenced by records of past climates. For us humans a hundred millennia is an infinite timespan, making the human-induced changes essentially permanent and irreversible. (That's one reason why drastic 'geoengineering' options are being considered. The safest such option, artificially pulling carbon dioxide back out of the atmosphere, is, however, far more expensive than going for suitable renewable[124] and nuclear[129] energy technologies, and carbon capture at source.[128]) Regarding sea levels 130 millennia ago, see for instance Overpeck, J. T., Otto-Bliesner, B. L., Miller, G. H., Muhs, D. R., Alley, R. B., and Kiehl, J. T., 2006: "Paleoclimatic evidence for future ice-sheet instability and rapid sea-level rise," *Science* **311**, 1747–1750, and references therein. Recently revised estimates appeared in Dyer, B., Austermann, J., D'Andrea, W. J., Creel, R. C., Sandstrom, M. R., Cashman, M., Rovere, A., and Raymo, M. E., 2021: "Sea-level trends across the Bahamas constrain peak last interglacial ice melt," *Proceedings of the National Academy of Sciences* **118**(33), e2026839118.

97. Farmer, J. D., *et al.*, 2019: "Sensitive intervention points in the post-carbon transition," *Science* **364**, 132–134. The authors remind us that the fossil fuel industry has long been heavily subsidized, to a "far greater" extent than renewable, low-carbon energy sources such as solar photovoltaics and wind. However, they also argue that there's hope of reaching a socio-economic tipping point, in the near future, now that renewables have become cheaper without subsidy. For further details see Ref. 124.

98. Shakun, J. D., Clark, P. U., He, F., Marcott, S. A., Mix, A. C., Liu, Z., Otto-Bliesner, B., Schmittner, A., and Bard, E, 2012: "Global warming preceded by increasing carbon dioxide concentrations during the last deglaciation," *Nature* **484**, 49–55.

99. Skinner, L. C., Waelbroeck, C., Scrivner, A. E., and Fallon, S. J., 2014: "Radiocarbon evidence for alternating northern and southern sources of ventilation of the deep Atlantic carbon pool during the last deglaciation," *Proceedings of the National Academy of Sciences* **111**, 5480–5484.

100. Marchitto, T. M., Lynch-Stieglitz, J., and Hemming, S. R., 2005: "Deep Pacific $CaCO_3$ compensation and glacial–interglacial atmospheric CO_2." *Earth and Planetary Science Letters* **231**, 317–336. This technical paper gives more detail on the role of sedimentary limestone sludge ($CaCO_3$) and seawater chemistry in the way carbon dioxide (CO_2) was stored in the deep oceans[95] during recent glacial cycles, and on the observational evidence. The evidence comes from meticulous and laborious measurements of tiny variations in chemicals that are important in the oceans' food chains, and in isotope ratios of various elements including oxygen and carbon, laid down in layer after layer of ocean sediments over many tens of millennia. Another reason for citing the paper, which requires the reader to have some specialist knowledge, is to highlight just how formidable the obstacles to building accurate models of the carbon cycle are, including for instance the way the plant-like *phytoplankton* pull carbon dioxide from the atmosphere, then die and carry carbon into the ocean depths. Such models try to represent oceanic carbon-dioxide storage along with observable carbon isotope ratios, which are affected by the way in which carbon isotopes are taken up by living organisms via processes of great complexity and variability. Not only are we far from modelling oceanic fluid-dynamical transport processes with sufficient accuracy, including turbulent eddies over a vast range of spatial scales, but we are even further from accurately modelling the vast array of biogeochemical processes involved throughout the oceanic and terrestrial biosphere — including for instance the biological adaptation and evolution of entire ecosystems and the rates at which the oceans receive mineral nutrients from rivers and airborne dust. The oceanic upper layers where phytoplankton live have yet to be modelled in fine enough detail to represent accurately the recycling of mineral nutrients simultaneously with gas exchange rates. It's fortunate indeed that we have the hard evidence, from ice cores,[122] for the atmospheric carbon dioxide concentrations that actually resulted from all this complexity.

101. Le Quéré, C., Rödenbeck, C., Buitenhuis, E. T., Conway, T. J., Langenfelds, R., Gomez, A., Labuschagne, C., Ramonet, M., Nakazawa, T., Metzi, N.,

Gillett, N., and Heimann, M., 2007: "Saturation of the Southern Ocean CO_2 sink due to recent climate change," *Science* **316**, 1735–1738. This work, based on careful observation, reveals yet another positive feedback that's increasing climate sensitivity to carbon dioxide injections.

102. Skinner, L. C., Primeau, F., Freeman, E., de la Fuente, M., Goodwin, P. A., Gottschalk, J., Huang, E., McCave, I. N., Noble, T. L., and Scrivner, A. E., 2017: "Radiocarbon constraints on the glacial ocean circulation and its impact on atmospheric CO_2," *Nature Communications* **8**, 16010.

103. Watson, A. J., Vallis, G. K., and Nikurashin, M., 2015: "Southern Ocean buoyancy forcing of ocean ventilation and glacial atmospheric CO_2," *Nature Geoscience* **8**, 861–864. This modelling study focuses on the amount of carbon dioxide injected from the deep ocean into the atmosphere during a deglaciation. In going from glacial to interglacial conditions, it's estimated that the contribution from ocean-circulation changes is roughly comparable, in order of magnitude, to that from changes in phytoplankton fertilization.

104. See for instance Abe-Ouchi, A., Saito, F., Kawamura, K., Raymo, M. E., Okuno, J., Takahashi, K., and Blatter, H., 2013: "Insolation-driven 100,000-year glacial cycles and hysteresis of ice-sheet volume," *Nature* **500**, 190–193. Importantly, their model includes a realistic, multi-millennial delay in the so-called 'isostatic rebound' or viscoelastic upward displacement of the Earth's crust as an overlying ice sheet melts, unloading the crust. Such a delay can keep the top surface of a massive ice sheet at lower, warmer altitudes long enough to enhance the melting effects of an insolation peak. The model uses an exponential time constant of 5 millennia for the isostatic rebound, a value whose order of magnitude is consistent with recent geophysical observations of the continuing isostatic rebound from the last deglaciation. The whole process is strongly 'nonlinear', as it's called technically, as distinct from a simple 'linear' response to the hundred-millennia components of orbital change, as used to be thought.

105. See for instance Schoof, C., 2010: "Ice-sheet acceleration driven by melt supply variability," *Nature* **468**, 803–806. This modelling study, motivated and cross-checked by recent observations of the accelerating ice streams in Greenland, is highly simplified but captures some aspects of subglacial meltwater flow networks, and their effects on the bulk motion of the ice sheet and ice streams within it. The rate of formation of meltwater is itself

increased by the 'biological darkening' due to the growth of meltwater-loving algae, as recently demonstrated on Greenland by the work of Professor Martyn Tranter and others.

106. See for instance Krawczynski, M. J., Behn, M. D., Das, S. B., and Joughin, I., 2009: "Constraints on the lake volume required for hydro-fracture through ice sheets," *Geophys. Res. Lett.* **36**, L10501. The standard elastic crack-propagation equations are used to describe the downward 'chiselling' of meltwater, which is denser than the surrounding ice, forcing a crevasse to open all the way to the bottom. This mechanism is key to the sudden drainage of small lakes of meltwater that accumulate on the top surface. Such drainage has been observed to happen within hours, for instance, on the Greenland ice sheet. The same 'chiselling' mechanism was key to the sudden breakup of successive portions of the Larsen ice shelf next to the Antarctic Peninsula, starting in the mid-1990s; see Figure 5 of Ref. 96 and, for instance, Banwell, A.F., MacAyeal, D.R., and Sergienko, O.V., 2013: "Breakup of the Larsen B ice shelf triggered by chain reaction drainage of supraglacial lakes," *Geophysical Research Letters* **40**, 5872–5876. Efforts to build more realistic ice-flow models include an important contribution from Pollard, D., DeConto, R.M., and Alley, R.B., 2015: "Potential Antarctic Ice Sheet retreat driven by hydrofracturing and ice cliff failure," *Earth and Planetary Science Letters* **412**, 112–121. Their model predicts about 3 metres of sea-level rise by the end of this century — yes, *three metres* — and far more in subsequent centuries.

107. Allen, Myles R., Frame, D. J., Huntingford, C., Jones, C. D., Lowe, J. A., Meinshausen, M., and Meinshausen, N., 2009: "Warming caused by cumulative carbon emissions towards the trillionth tonne," *Nature* **458**, 1163–1166. The trillionth tonne of carbon emitted into the atmosphere is, by some estimates, the threshold for climate change to be kept within acceptable bounds. Tipping-point concerns and recent weather extremes now suggest a lower threshold; see for instance McIntyre, M. E., 2023: "Climate tipping points: A personal view," *Physics Today* **76**(3) 44–49, and Kemp, L., Xu, C., Depledge, J., Ebi, K. L., Gibbins, G., Kohler, T. A., Rockstrom, J., Scheffer, M., Schellnhuber, H. J., Steffen, W., Lenton, T. M., 2022: "Climate endgame: Exploring catastrophic climate change scenarios," *Proceedings of the National Academy of Sciences* **119**(34), e2108146119.

108. Valero, A., Agudelo, A., and Valero, A., 2011: "The crepuscular planet: a model for the exhausted atmosphere and hydrosphere," *Energy* **36**, 3745–3753. This careful discussion lists the amounts of proven and estimated fossil-fuel reserves including coal, oil, gas, tar sands, and clathrates, showing that if burnt, they would produce emissions vastly greater than a trillion tonnes of carbon.[107] The amount of methane in clathrates is estimated to be of the order of, respectively, 40 and 100 times that in proven reserves of shale gas and conventional natural gas.

109. See for instance Shakhova, N., Semiletov, I., Leifer, I., Sergienko, V., Salyuk, A., Kosmach, D., Chernykh, D., Stubbs, C., Nicolsky, D., Tumskoy, V., and Gustafsson, O., 2014: "Ebullition and storm-induced methane release from the East Siberian Arctic Shelf," *Nature Geoscience* **7**, 64–70. As regards current methane leakage rates from fossil-fuel infrastructure, see for instance Lauvaux, T., Giron, C., Mazzolini, M., d'Aspremont, A., Duren, R., Cusworth, D., Shindell, D., and Ciais, P., 2022: "Global assessment of oil and gas methane ultra-emitters," *Science* **375**, 557–561.

110. See also Andreassen, K., Hubbard, A., Winsborrow, M., Patton, H., Vadakkepuliyambatta, S., Plaza-Faverola, A., Gudlaugsson, E., Serov, P., Deryabin, A., Mattingsdal, R., Mienert, J., and Bünz, S, 2017: "Massive blow-out craters formed by hydrate-controlled methane expulsion from the Arctic seafloor," *Science* **356**, 948–953. It seems that some of the clathrates in high latitudes have been melting ever since the later part of the last deglaciation, probably contributing yet another positive feedback, both then and now. Today, the melting rate is accelerating to an extent that hasn't been well quantified but is related to ocean warming and to the accelerated melting of the Greenland[105, 106] and West Antarctic ice sheets, progressively unloading the strata beneath. Reduced pressures lower the clathrate melting point.

111. See for instance Cramwinckel, M., Huber, M., Kocken, I. J., Agnini, C., Bijl, P. K., Bohaty, S. M., Frieling, J., Goldner, A., Hilgen, F. J., Kip, E. L., Peterse, F., van der Ploeg, R., Röhl, U., Schouten, S., and Sluijs, A., 2018: "Synchronous tropical and polar temperature evolution in the Eocene," *Nature* **559**, 382–386. This recent data study from ocean sediment cores brings many lines of evidence together, confirming earlier conclusions that the hottest prolonged period was roughly between 53–50 million years ago, peaking around 52–51 million years ago except for a relatively

brief, extremely hot 'Palaeocene–Eocene Thermal Maximum' (PETM)[112] at the start of the Eocene around 56 million years ago. Also presented are improved estimates of tropical sea surface temperatures, with maxima of the order of 35°C around 52 million years ago and nearly 38°C during the PETM. Tropical sea surface temperatures today are still mostly below 30°C.

112. See for instance Giusberti, L., Boscolo Galazzo, F., and Thomas, E., 2016: "Variability in climate and productivity during the Paleocene–Eocene Thermal Maximum in the western Tethys (Forada section)," *Climate of the Past* **12**, 213–240. The early Eocene began around 56 million years ago with the Palaeocene–Eocene Thermal Maximum (PETM), a huge global warming episode with accompanying mass extinctions now under intensive study by geologists and palaeoclimatologists. It required a massive increase in atmospheric carbon-dioxide concentration, which probably came from volcanism added to by clathrate melting, the methane from which was then converted to carbon dioxide. Peatland burning might also have contributed. The western Tethys Ocean was a deep-ocean site at the time and so provides biological and isotopic evidence both from surface and from deep-water organisms, such as foraminifera with their sub-millimetre-sized carbonate shells. The accompanying increase in the extremes of storminess is evidenced by massive soil erosion from what the authors describe as "storm flood events".

113. See for instance Thewissen, J. G. M., Sensor, J. D., Clementz, M. T., and Bajpai, S., 2011: "Evolution of dental wear and diet during the origin of whales," *Paleobiology* **37**, 655–669. Today's hippopotami are genetically close to today's whales, dolphins, and other cetaceans.

114. Foukal, P., Fröhlich, C., Spruit, H., and Wigley, T. M. L., 2006: "Variations in solar luminosity and their effect on the Earth's climate," *Nature* **443**, 161–166. An extremely clear review of some robust and penetrating insights into the relevant solar physics, based on a long pedigree of work going back to 1977. For a sample of the high sophistication that's been reached in constraining solar models, see also Rosenthal, C. S. *et al.*, 1999: "Convective contributions to the frequency of solar oscillations," *Astronomy and Astrophysics* **351**, 689–700, and more recently Buldgen, G. Salmon, S., and Noels, A., 2019: "Progress in global helioseismology: A new light on the solar modeling problem and its implications for solar-like stars," *Frontiers in Astronomy and Space Sciences* **6**, Article 42.

115. Solanki, S. K., Krivova, N. A., and Haigh, J. D., 2013: "Solar irradiance variability and climate," *Annual Review of Astronomy and Astrophysics* **51**, 311–351. This review summarizes and clearly explains the recent major advances in our understanding of radiation from the Sun's surface, showing in particular that its magnetically induced variation cannot compete with the carbon-dioxide injections I've been talking about.

116. Cyclonic storms, with or without embedded thunderstorms, transport heat and weather fuel poleward. For evidence that this process took place in Eocene times, as it does in ours, see for instance Carmichael, M. J., Inglis, G. N., Badger, M. P. S., Naafs, B. D. A., Behrooz, L., Remmelzwahl, S., Monteiro, F. M., Rohrssen, M., Farnsworth, A., Buss, H. L., Dickson, A. J., Valdes, P. J., Lunt, D. J., and Pancost, R. D., 2017: "Hydrological and associated biogeochemical consequences of rapid global warming during the Palaeocene–Eocene Thermal Maximum,"[112] *Global and Planetary Change* **157**, 114–138. See for instance their Figure 8, showing the effects of weather-fuel transport from the tropics into middle and high latitudes.

117. Saravanan, R., 2022: *The Climate Demon: Past, Present, and Future of Climate Prediction*, Cambridge University Press. This very readable book gives an up-to-date view of the state of the art in climate modelling, and of the different kinds of uncertainty involved, paying close attention to the chaotic nature of weather and climate dynamics and its relation to chaos theory. The author is a respected expert in the field.

118. Palmer, T. N., 2022: *The Primacy of Doubt: From Climate Change to Quantum Physics, How the Science of Uncertainty Can Help Predict and Understand Our Chaotic World*, Basic Books and Oxford University Press. This book gives another up-to-date expert view of climate modelling and chaos theory, complementing Ref. 117. A possible future development in modelling is to use new 'noisy computer hardware' that deliberately exploits quantum noise in the circuitry, and reduces arithmetical precision, to economize on power consumption while making partial allowance for the chaotic, in a sense noisy, dynamics of weather and climate. In that way it supplements an important modelling technique called 'noisy parametrization' or 'stochastic parametrization'. And beyond climate issues, the book, in a remarkable way, goes on to discuss how chaos theory might help to resolve the profound enigmas at the foundations of quantum theory.[94]

119. Kendon, E. J., Roberts, N. M., Fowler, H. J., Roberts, M. J., Chan, S. C., and Senior, C. A., 2014: "Heavier summer downpours with climate change revealed by weather forecast resolution model," *Nature Climate Change* **4**, 570–576. See also, for instance, Knote, C., Heinemann, G., and Rockel, B., 2010: "Changes in weather extremes: assessment of return values using high resolution climate simulations at convection-resolving scale," *Met. Zeitschr.* **19**, 11–23, and Mahoney, K., Alexander, M. A., Thompson, G., Barsugli, J. J., and Scott, J. D., 2012: "Changes in hail and flood risk in high-resolution simulations over Colorado's mountains," *Nature Climate Change* **2**, 125–131.

120. See for instance Emanuel, K. A., 2005: "Increasing destructiveness of tropical cyclones over the past 30 years," *Nature* **436**, 686–688. For tropical cyclones (hurricanes, typhoons), the simple picture of a fast runup to peak intensity does not apply. Intensification is a relatively slow and complex process. Tropical cyclones scoop up most of their weather fuel directly and continually from the sea surface, making them sensitive to local upper-ocean heat content and sea-surface temperature. They are sensitive also to conditions in their large-scale surroundings including temperatures at high altitudes, which in turn are affected by complex cloud-radiation interactions, and atmospheric haze or aerosol. Our current modelling capabilities fall far short of giving us a complete picture. The author of the cited paper, Professor Kerry Emanuel, is a Foreign Member of the Royal Society and a leading expert on these matters. He tells me that the conclusions in the cited paper now need modification for the Atlantic because of atmospheric aerosol issues not then taken into account. However, the conclusions for the Pacific still appear valid (personal communication, 2020). Those conclusions still suggest a tendency for the models to err on the side of underprediction, rather than overprediction, of future extremes of storminess.

121. Abram, N. J., Wolff, E. W., and Curran, M. A. J., 2013: "Review of sea ice proxy information from polar ice cores," *Quaternary Science Reviews* **79**, 168–183. The light graph near the top of Figure 20 above comes from measuring the concentration of sea salt in the Dronning Maud Land ice core from East Antarctica. The review carefully discusses why this measurement should correlate with the area of sea ice surrounding the continent, as a result of long-range transport of airborne, sea-salt-bearing powder snow

blown off the surface of the sea ice. The graph shows sea-salt concentration divided by the estimated time of ice-core-forming snow accumulation in the ice-core layer measured — hence the label 'flux', or rate of arrival, rather than 'concentration'.

122. A quick overview of the cross-checking of the ice-core atmospheric carbon dioxide data can be found, together with research-literature references, at https://doi.org/10.5281/zenodo.7573126. It was written by a leading ice-core expert, Eric Wolff FRS, in response to criticism from another colleague.

123. See for instance Shackleton, N. J., 2000: "The 100,000-year ice-age cycle identified and found to lag temperature, carbon dioxide, and orbital eccentricity," *Science* **289**, 1897–1902. See Figure 4B. The late Sir Nick Shackleton FRS was a respected expert on palaeoclimatic data and its interpretation, especially the data from ocean sediment cores.

124. Ref. 97 argues from the history of sociological tipping points that we are indeed closer, now, to finding the political will to take climate seriously. Despite the continued pressures to maintain fossil-fuel subsidies and to build new oil wells and gas wells, countervailing pressures have been increasing. In part that's because, as the authors point out, "renewable energy sources such as solar photovoltaics and wind have experienced rapid, persistent cost declines" — and are available for use with smart grids, dynamic electricity pricing, large-scale battery storage, and green hydrogen production — whereas, despite "far greater investment and subsidies, fossil fuel costs have stayed within an order of magnitude for a century." That's a new economic reality, of which for instance the state of South Australia has already taken advantage, having transitioned from coal to renewables and battery storage in just a few years, with a net reduction in the cost of electricity. And the schoolchildren's and other mass movements suggest that a sociological tipping point is indeed closer now that young people, especially, are making their voices heard more and more clearly.

125. Isaacson, W., 2021: *The Code Breaker: Jennifer Doudna, Gene Editing, and the Future of the Human Race*, Simon & Schuster. This book by Walter Isaacson describes the rapidity with which molecular biologists, using new techniques including CRISPR-Cas9 gene editing, have been building revolutionary new therapies and new, precisely tailored vaccines including the inherently safe, simple, and versatile MRNA-based vaccines against the

virus that causes COVID-19. (The MRNA, messenger ribonucleic acid — the vaccines' only active ingredient — directly instructs cells to make a specified protein, which can be tailored to match a protein on the outer surface of the virus, making it a target for the immune system.) At the centre of the CRISPR-Cas9 story was the work of Nobel laureate Jennifer Doudna, who has also led discussions of the far-reaching ethical issues that arise. Doudna describes having had a nightmare in which Adolf Hitler asked her to teach him how to do gene editing. There's also the nightmare of unleashing ineptly tailored 'gene drives' into ecosystems.

126. See for instance King, D., Schrag, D., Zhou, D., Qi, Y., Ghosh, A., and co-authors, 2015: "Climate Change: A Risk Assessment," Cambridge Centre for Science and Policy. See also the relatively brief RS-NAS 2014 (UK Royal Society and US National Academy of Sciences) report "Climate Change: Evidence & Causes," and its 2020 update with the same title. These reports and their successors come from high-powered teams of scientists, supplementing the vast IPCC reports and emphasizing the careful thinking that's been done.

127. Stern, N., 2009: *A Blueprint for a Safer Planet: How to Manage Climate Change and Create a New Era of Progress and Prosperity*, London, Bodley Head. Even as early as 2009, the eminent economist Nicholas Stern was pointing out that the path to a green economy is a path to renewed prosperity.

128. Oxburgh, R., 2016: "Lowest Cost Decarbonisation for the UK: The Critical Role of CCS". Report to the Secretary of State for Business, Energy, and Industrial Strategy from the Parliamentary Advisory Group on Carbon Capture and Storage, September 2016.

129. Stein, P., 2020: Interview with BBC News on 24 January 2020 on the Rolls Royce consortium, a project to develop and build 'small modular reactors' to produce nuclear power. Paul Stein, the Chief Technology Officer at Rolls Royce, said that the design and costing are already complete and that both have been scrutinized by the Royal Academy of Engineering and by the Treasury of the UK government. He argued persuasively that basing the design on advanced but conservative engineering and manufacturing techniques, more like car manufacture, will drive costs down — by contrast with large, one-off civil construction projects like the UK's Hinkley Point C nuclear power station, for which "history shows that costs go up with time".

130. McIntyre, M. E., 1998: "Lucidity and science: III. Hypercredulity, quantum mechanics, and scientific truth." *Interdisciplinary Science Reviews* **23**, 29–70. See the *Corrigendum* to Part III, a slightly corrupted version of which was published in the December 1998 issue of *Interdisciplinary Science Reviews*. Further discussion, a precursor to this book, is in McIntyre, M. E., 2000: "Lucidity, science, and the arts: what we can learn from the way perception works." *Bull. Faculty Human Devel.* (Kobe University, Japan), 7(3), 1–52. (Invited keynote lecture to the 4th Symposium on Human Development, *Networking of Human Intelligence: Its Possibility and Strategy*.)

Index